Harold J. Rosen and Tom Heineman
CONSTRUCTION SPECIFICATIONS WRITING: PRINCIPLES AND
PROCEDURES, Third Edition

Leo Diamant and C. R. Tumblin
CONSTRUCTION COST ESTIMATES, Second Edition

Thomas C. Schleifer
CONSTRUCTION CONTRACTORS' SURVIVAL GUIDE

A. C. Houlsby
CONSTRUCTION AND DESIGN OF CEMENT GROUTING: A
GUIDE TO GROUTING IN ROCK FOUNDATIONS

David A. Day and Neal B. H. Benjamin
CONSTRUCTION EQUIPMENT GUIDE, Second Edition

Terry T. McFadden and F. Lawrence Bennett
CONSTRUCTION IN COLD REGIONS: A GUIDE FOR PLANNERS,
ENGINEERS, CONTRACTORS, AND MANAGERS

S. Peter Volpe and Peter J. Volpe
CONSTRUCTION BUSINESS MANAGEMENT

CONSTRUCTION BUSINESS MANAGEMENT

CONSTRUCTION BUSINESS MANAGEMENT

S. PETER VOLPE, *President*
PETER J. VOLPE
Volpe Construction Co.
Malden, Massachusetts

A WILEY-INTERSCIENCE PUBLICATION

JOHN WILEY & SONS, INC.

New York • Chichester • Brisbane • Toronto • Singapore

Please note that the masculine pronoun has been used
throughout this book as a matter of convenience only. It is
recognized that both sexes can and do function as construction
workers.

In recognition of the importance of preserving what has been
written, it is a policy of John Wiley & Sons, Inc., to have books
of enduring value published in the United States printed on
acid-free paper, and we exert our best efforts to that end.

Library of Congress Cataloging in Publication Data:
Volpe, S. Peter (Sonnino Peter), 1917-
    Construction business management / by S. Peter Volpe and Peter J.
Volpe.
        p. cm. — (Wiley series of practical construction guides)
    Includes index.
    ISBN 0-471-53636-9
    1. Construction industry—Management. I. Volpe, Peter J.
II. Title. III. Series.
HD9715.A2V62 1991   91-22689
624'.068—dc20   CIP

Printed in the United States of America

10 9 8 7 6 5 4 3 2 1

SERIES PREFACE

Congratulations! You've just bought a profit-making tool that is inexpensive and requires no maintenance, no overhead, and no amortization. Actually, it will increase in value for you each time you use this volume in the Wiley Series of Practical Construction Guides. This book should contribute toward getting your project done under budget, ahead of schedule, and out of court.

For nearly a quarter of a century, over 50 books on various aspects of construction and contracting have appeared in this series. If one is still valid, it is "updated" to stay on the cutting edge. If it ceases to serve, it goes out of print. Thus you get the most advanced construction practice and technology information available from experts who use it on the job.

The Associated General Contractors of America (AGC) statistician advises that the construction industry now represents close to 10% of the gross national product (GNP), some 410 billion dollars worth per year. Therefore, simple, off-the-shelf books won't work. The construction industry is unique in that it is the only one where the factory goes out to the buyer at the point of sale. The constructor takes more than the normal risk in operating a needed service business.

Until the advent of the series, various single books (many by professors), magazine articles, and vendors' literature constituted the total source of information for builders. To fill this need, this series has provided solid usable information and data for and by working constructors. This has increased the contractors' earning capacity while giving the owner a better product. Profit is not a dirty word. The Wiley Series of Practical Construction Guides is dedicated to that cause.

M. D. MORRIS, P.E.

Ithaca, New York
November 1989

v

PREFACE

The construction industry has come a long way from its early beginnings of people who dug caves into the mountainside for shelter. Today we have stretched what we thought were the limits even five or ten years ago. Buildings are reaching higher and higher every day. New and stronger and lighter materials are still being developed almost daily. Innovations in heating, cooling, and electrical systems including the addition of computerized operation are enough to make one's head swim. New innovations occur every day making us wonder where we are headed. Is it any wonder that people outside our industry have such a hard time understanding our business?

Although this book does not go into these mind-boggling developments it does go into the basic understanding and workings of our business. It details how a company can operate in today's environment.

Ours is a business that can be very rewarding and yet at times can be very frustrating and even disastrous. The sense of accomplishment when a difficult project is finished is very strong. You dig a hole in the ground and, like planting a seed, you watch this edifice grow from the foundations to the skeletal frame, then the skin application to the interior finish. There stands a structure that is *your* accomplishment, a seed you nurtured from a set of plans to a real building. The structure may have far-reaching effects and only time will tell its real value.

We hope the contents of this book will help those who follow us become better constructors. It is our wish that professionalism will become more of a factor in our industry. This book purposely describes our ideas in a simple and direct manner. We believe that those outside our industry, such as Owners, Bankers, Students, and Insurance people, can gain a valuable insight, to our business. Architects and Engineers may also find some information valuable to them. It is our hope that who ever the readers, they will gather some useful ideas from the book, which will make us feel it was worth writing.

We gratefully acknowledge the help of Dan Morris P. E. of the Wiley Series of Practical Construction Guides, who encouraged us in planning and writing this book. We thank our wives, Armita and Linda, for their forbearance in the time we took from our normal activities to write this book.

S. PETER VOLPE
PETER J. VOLPE

Malden, Massachusetts
August 1991

PREFACE TO CONSTRUCTION MANAGEMENT PRACTICE

Construction is an industry that remains a mystery to some. By sharing the experience and knowledge I have acquired over the past 35 years it is my hope that each reader will gain a better insight into the industry. Outsiders can learn how a contractor organizes and controls his work. Contractors will discover what seems to me to be the advantage of increasing our level of professionalism in daily business activities.

Contracting is rewarding in many ways. You feel a great sense of accomplishment as a structure rises from a hole in the ground to a solid, permanent feature of the landscape. Your influence in the community, for better or worse, is far reaching. You work with a wide range of people every day. The only thing that is constant for a contractor is change.

It is my wish that the methods of operation described will be of benefit to those in our industry who are looking for some guidance in their business. Some who are looking for methods of accomplishing certain phases of their operation may find a few helpful ideas. The book could be useful not only to new or growing firms but also to anyone associated with construction—banker, insurance man, architect, engineer, or owner.

The style of presentation is purposely simple and direct. I have used the first person wherever it seemed to be more natural. If this book gives you at least one good idea that helps your work, I shall consider it a success.

Grateful acknowledgment is given to Dan Morris for encouragement and suggestions in planning this book.

S. PETER VOLPE

Malden, Massachusetts
February 1971

CONTENTS

GLOSSARY

A.G.C.A.	Associated General Contractors of America
A.I.A.	American Institute of Architects
AS BUILT DRAWINGS	Original drawings modified to show actual installation information.
BID DAY	The day on which bids are received by the Owner.
BONDING CAPACITY	The total dollar value of contracts that the surety company will guarantee for a contractor.
CHANGE ORDERS	A written authorization to perform either more or less work than originally specified.
CHIEF ESTIMATOR	The person in charge of the Estimating Department.
CHIEF PROJECT MANAGER	The person who oversees the projects and the project managers.
COMPUTERIZATION	Utilization of computers for the many functions of your business.
CONTINGENCY	A sum of money added to an estimate to cover unforeseen items when drawings are not fully completed. It will also cover anticipated increases of labor and material costs on a long term contract.
C.P.A.	Certified public accountant.
C.P.M.	A scheduling method known as "Critical Path Method" to control time and sometimes costs of a project.
C.S.I.	Construction Specification Institute
CUBIC FOOT	A measurement of quantity 12 inches long, 12 inches wide, and 12 inches high.

CUBIC YARD	A quantity measure of 27 cubic feet.
COST KEEPING SYSTEM	A system of keeping costs, of both labor and materials which will enable you to know how the costs compare with the estimate.
ENGINEER	The person who generally performs the structural, mechanical, or electrical design for a project.
ESTIMATE	The total assembled cost of labor, materials, overhead, and profit for the project. This can also be called the bid.
ESTIMATOR	The person on your staff who translates the information from the plans and specifications to quantities that are used to make up the estimate.
FORMWORK	The temporary support built in the shape needed to hold poured concrete until it has reached sufficient strength to stand on its own. This formwork may be built of wood, steel or other material strong enough to support the concrete.
GENERAL CONDITIONS	Usually a standard document that describes the responsibilities of the parties to the contract.
INSURANCE CERTIFICATE	A document that gives proof of insurance coverage (in lieu of the policy) it names the parties of interest and the dollar limits of the coverages.
INSURANCE CHECK LIST	A list of insurance coverages available as a check for coverages needed on a project.
JOB MEETINGS	Meetings held at the job site periodically to discuss progress and review any problems that have to be resolved.
JOINT VENTURES	A partnership formed by two or more contractors usually to bid on larger jobs that a single contractor may not be able to bond.
JURISDICTIONAL DISPUTE	A condition where two or more craft unions may claim the same item of work. A strike may be threatened and may actually occur if the dispute is unresolved.
LINE OF CREDIT	The limit a bank will lend you at any given time. It is also referred to as your borrowing capacity.
MULTIPLE CONTRACT PROJECTS	A project that has several prime contractors without any one of them being in charge.
PAYMENT REQUISITIONS	A form that lists the items of work completed during a given time frame, such as monthly, which becomes the basis of your being paid for the work completed.

PERFORMANCE AND PAYMENT BONDS A bond furnished by a surety company guarantying that the contractor will finish the project and pay all the outstanding bills.

PLANS Drawings for a project are called plans or prints. They are drawn to scale and include details and the critical dimensions. The contractor translates this information into a finished structure.

PROJECT MEETING MINUTES The minutes of the job meetings held listing the items discussed and the actions to be taken by the various parties involved.

PUNCH LIST A list of minor deficiencies to be corrected to complete the project, usually prepared by the architect at the time of substantial completion.

PURCHASE ORDER A form used to confirm a purchase.

QUANTITY TAKEOFF A list of the quantities of material required for a project.

REINFORCED CONCRETE Concrete that contains reinforcing steel rods placed so as to give the concrete the strength needed to support the structure. They are also called "rebars".

S.A.E. Society of Automotive Engineers

SAFETY PROGRAM The outline of company policy regarding safety.

SHOP DRAWINGS A detailed plan of how an item is to be fabricated or installed.

SPECIAL CONDITIONS A supplemental set of conditions that modifies or adds to the general conditions.

SPECIFICATIONS A book of written details for the materials and construction procedures for the project. It is prepared by the architect and engineers and is generally broken down into sections following the C.S.I. format. These are used in conjunction with the plans to form the basis of the work to be done.

SQUARE FOOT A surface area of 12 inches square.

SUBCONTRACT A contract awarded to a specialty contractor for a specific item of work such as Mechanical, Electrical, Flooring, etc. These specialty contractors are known as "Trade Contractors", "Subcontractors" or "Subs".

SUBSTANTIAL COMPLETION The point of completion of the project when the Owner can beneficially occupy the structure. The remaining items of work generally consist of the punch list.

FIGURES SHOWN IN THE BOOK

CONSTRUCTION BUSINESS MANAGEMENT

CHAPTER 1

Starting Out

To be a successful contractor requires two basic skills: first, one must know how to estimate work so bids will win contracts; second, work must be completed within the cost and time estimates—this is the "make or break" of a firm. Like it or not, the objective of any business venture is profit.

Construction is often thought to be a highly profitable business, but in reality it is highly competitive. Profit margins can be small. Profit is not a dirty word; it is what makes people want to go into business. Construction is an interesting and exciting business. Although there is much routine work, the danger of the unexpected is always present.

Attitude of management in any business is a most important consideration. Attitude shows in every phase of your work. A positive attitude is easily identified by employees and customers alike. Managing a business requires full-time attention. This may sound elementary, but many contractors, having met with mild success, begin to think that the business will run by itself. There are so many risks, variables, and unforeseen conditions that a wrong decision, or no decision, by anyone in the firm can result in financial disaster.

Planning and scheduling the details of work cannot be ignored. Nor can they be worked out each morning at the job site. Advance planning is necessary. Depending on the complexity of the project, advance planning can be measured in hours or days and yes, even months. Availability of specified materials or manpower will govern the time required. Although some details are more important than others, they all tie into an end result that could spell success or failure for your firm.

We believe that all firms should establish a good set of principles, then follow them in both good and bad times.

The Associated General Contractors of America (A.G.C.A.) has as its motto, "Skill, Integrity, and Responsibility." These three words describe what is expected of all AGCA members. We see no circumstance where there should be any deviation from this ideal.

Here is our interpretation of the motto:

Skill: The "know-how" to do any job you undertake. When a firm lacks experience in a particular skill, it should hire a specialist or consultant to help do the job properly and solve any problem if it occurs.

Integrity: Fairness and honesty in all your dealings with clients, architects, subcontractors, material suppliers, and employees. Do not let the temptation of a quick profit ruin your reputation, your business, and your life. The incident can overshadow years of good work. The integrity of the firm should be preserved at all costs.

Responsibility: Carrying out all provisions of the contract. Upon accepting a contract, responsibilities are assumed for undertaking and providing an expected end result. Nothing less must be provided. There is a saying, "Do it right and do it once, do it wrong and do it twice." The second time is very expensive because both pocketbooks and reputations suffer.

A lack of integrity and responsibility will destroy image and the business faster than you may realize. A lack of skill may take a bit longer.

1.1 HOW TO GET BUSINESS

There are many ways to solicit business in the construction industry. The three most common ways are: (1) bidding on public work against all bidders; (2) bidding on private work against a selected list of bidders; and, (3) negotiating work. Most contractors start out in business fighting in the competitive mill of bidding on work against all bidders. By doing a good job and showing skill, integrity, and responsibility, a contractor can graduate to the level of selected bidder lists, then on to negotiated work.

Public work is highly competitive. The law requires that contracts be awarded to the low bidder, providing he meets specified basic qualifications. The bidders list is usually long, and it often seems that the contractor who makes a mistake in his estimate is the low bidder! However, public work provides plenty of experience plus the opportunity to deal with architects and engineers, many of whom also do private work.

Some architects do both public and private work. They have quite a bit of influence when making up a list of contractors selected to bid on projects. If your firm has shown skill and good work on any project, working successfully with an architect, it is much easier getting on that architect's list of bidders. In some cases if exceptional ability has been shown, you may even be considered as the contractor on a negotiated basis contract.

Private work can also be highly competitive. However, owners can award the contract to anyone they choose. To simplify the procedure, many owners limit bidding to a selected list of contractors. These lists are based on reputation and a record of successful on-time completion of similar projects. Quality and on—time completion are frequently key factors in private work. Generally, when bids are taken from a selected lists of bidders, the low bidder gets the job. Generally all bidders have been prequalified. As a result, theoretically, all the bidders are acceptable.

The owners select firms so they do not have to worry about who is the low bidder. If they did not prequalify, anyone could bid and they would be evaluating the contractors first and not the price.

Negotiated work is generally profitable and not usually so competitive. Owners expect quality work and want the better contractors. Negotiated contracts can be for a fixed price (usually called "lump sum"), cost of work plus a fee, guaranteed maximum price, or other mutually satisfactory agreements. All negotiated work requires a high level of mutual trust. When an owner decides to negotiate the work, he has to have confidence in a contractor's ability. The contractor works with the architect/engineer from the inception of the project. This requires a good knowledge of costs, construction methods and availability of materials to help make the right decisions to ensure the viability of the project. The contractor must have the ability to provide accurate budgets from programs and/or design development drawings. As the drawings progress he is expected to update the estimate and if the budget changes keep the owners informed. They must know why the budget goes up or down and may ask you to suggest changes to keep the project within the original budget. This means using different materials that are more economical but still of good quality. Suggesting other construction methods that save money is another item that must also be considered. You must have the ability to respond to this kind of a situation to be effective. This is called "value engineering," which is important in keeping a project within its budget.

1.2 HOW YOUR FUTURE IS SHAPED

Many contracting firms start business in a slap-bang style, but sooner or later get the desire to do a monumental type of job. Getting on the bid list is greatly influenced by reputation. It is amazing how well clients remember both their good and their bad experiences. If the problem is not the company's fault, you must be able to prove this to the owner and architect at the time. It is difficult to correct this impression long after the fact. Documentation is very important in these instances.

Contractors involved in trouble of their own creation are seldom invited to bid again for the same owner. They end up back in the "slug-it-out" competitive market. Top management must take the initiative to stay among the selected bidders. This attitude should be projected from the top down so the small job problems and details do not become a steady matter of irritation and controversy. Often small problems rather than large ones tarnish your reputation with the owners. All problems must be attended to without delay.

If people working on the project do not have the authority to make hard decisions, this could be the cause of some of the problems. If they do not have the power, then it had better be given to them or someone with the authority to act should be put in charge. Sometimes decisions have to be compromised because the problem may involve some gray areas. You must be able to judge the degree of the gray area, and how much effort should be contributed to solve the problem.

These types of statesmanlike actions show your willingness and ability to work out problems, thereby enhancing your reputation.

1.3 HOW YOUR ATTITUDE IS PROJECTED

Dealings with the client, the architect, subcontractor, material supplier, and labor are never hidden. Sooner or later everyone from your executive vice-president to your laborers will form a reasonably accurate opinion of your business ethics. To be honest is the only way to do business. Any compromise in this attitude is not conducive to long-term success. Following your example and instructions, employees will adhere to the principles of good workmanship, honest dealings, and good practices.

An important factor to owners and architects is your relations with employees of the firm. It is important for a firm to have a stable work force. If your employees have been with the firm for a good period of time it is interpreted by owners and architects to mean that they are treated fairly, which would indicate to them that they too would be treated the same. It also means that they will have continuity of personnel on that project. Turnover of personnel on a project can disrupt the job with new people coming in and having to learn what has been going on and trying to catch up with the project details. The same holds true in community relations by having the same people represent the company year in and year out.

Actions outside your firm and in the community also help shape your attitude; they show that you are people with whom others will want to be associated with and therefore be willing to work with. People like to do business with successful people.

1.4 KEEPING GOOD SUBCONTRACTOR RELATIONS

Honesty pays further dividends with your subcontractors. They like to do business with people who treat them well, pay their bills on time, and do not back-charge them unfairly. The subcontractor makes a better profit on a well-run job that proceeds on schedule. It is important that you direct the job. Do not just let it run day to day on its own momentum. The job has to be planned and scheduled so subcontractors can plan and schedule their work and people. If subcontractors are requested to increase forces one week, slow down the next, and back up later, it will cost them money. This planning is very important. Good planning results in subcontractors giving you lower prices on future work. This makes your own future bids more competitive.

Money is important to you, but it is also important to the subcontractors. Requisitions should be processed promptly so payments are on time. Prompt payment means a lot to subcontractors. They like to pay their bills on time and earn a discount. This too, makes a subcontractor want to give you better prices, because he can plan his own timely payments. Every area of business activity is important and needs careful attention from you and your employees.

1.5 YOUR PUBLIC RELATIONS

Public relations help create a favorable image in the community and areas of operation. You can make yourself known in various ways, deriving pleasure and satisfaction at the same time. Active participation in a contractors's association will help upgrade the construction industry. Learning how others solve current problems can also be a bonus. The Chamber of Commerce and such service clubs as Kiwanis, Rotary, and Lions will broaden company contracts. Church activities and charitable work develop skills working with all types of people. Local municipal affairs or political office offer another challenge.

Because of your enthusiasm, effort, and ability in these outside areas, people you touch will realize that the firm must be competent and trustworthy and will recommend the company without hesitation. An introduction by a mutual friend could be extremely valuable in some later business situation.

A well illustrated brochure will show a firm's capabilities, background and experience. Do not underestimate its value. Use the services of a good public relations firm to ensure an effective professional result. However, although advertising can be useful, it will never take the place of personal effort in creating a favorable image.

1.6 KEEPING A GOOD IMAGE

Once you have established yourself, do a good job every time and your reputation will continue to improve. Encourage your employees to follow your own example in some sort of public service activity. You and your firm will be a valued asset to your community.

CHAPTER 2

Your Organization

2.1 NEEDS

The successful construction firm has well balanced skills in both field and office operations. Often the founder of a construction firm is a field person who managed jobs with skill and imagination. Upon completion of a profitable project and receipt of a healthy bonus, he starts out on his own. But because the office functions of estimating and accounting were handled smoothly and efficiently by others in previous employment, he may underrate their importance. On the other hand, the inside man may feel that he can always hire a skilled field man if he starts out and, depending on his capability, get the job done at a profit. This could also turn into a disaster unless watched carefully.

In actual practice, both field and office operations must be properly blended for success and growth. Analysis has shown the major cause of construction firm failures to be inexperience. This does not mean that you must be a genius or have the full range of experience to start a business. Just recognize the limits of your knowledge and hire or join with a person to complement your own abilities.

Many new firms in earlier days were formed by people from the trades. Today, however, more are formed by construction people who come from the engineering ranks or those who have a construction or business oriented degree as part of their experience. Even more superintendents are coming from engineering and construction education oriented schools.

2.2 SMALLER FIRM

A partnership made up of a good superintendent and an experienced estimator is frequently the beginning for a small firm. The elements for success are here. If trouble

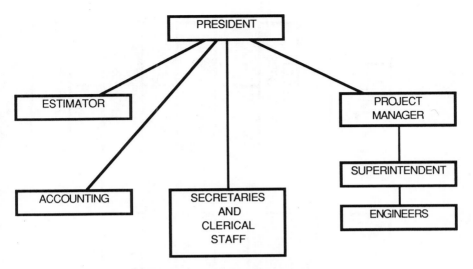

FIGURE 2.2 Organization Chart—Smaller Job

occurs, it probably is a result of a personality clash or inattention to the details of the business.

Figure 2.2 shows in a simplified manner the organization of a small construction firm. Whether your firm is small or large, the same functions are needed. In the smaller firms one person performs two or three functions. In a very small firm one person may well perform all functions. The point is to recognize that all functions must be performed one way or another and that each should be assigned to a specific individual.

2.3 LARGER FIRM

Figure 2.3 shows a typical organization in a larger firm. The title of president designates the operating head, who could be the sole owner or senior partner. The size of the firm is determined by the annual dollar volume of work put in place.

2.3.1 President

The typical head man has a well-rounded background. He has worked his way through the various aspects of construction and has experience in both field and office. His assistants include a field man, a chief estimator, and an accountant. He is free to look for new work by visiting owners, architects, engineers, and developers. He helps resolve any serious problems that occur. He ensures that estimates are prepared properly and on time. He sees that projects are kept on schedule and budgets are met. He follows through to make sure the firm is making a profit.

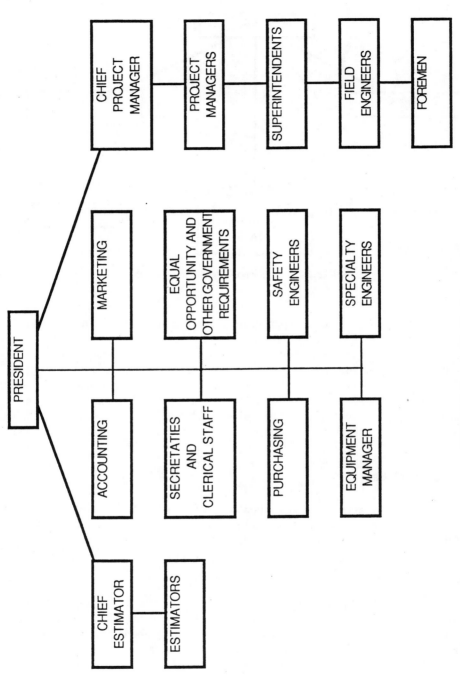

FIGURE 2.3 Organization Chart—Larger Job

2.3.2 Chief Estimator

The chief estimator helps select projects to be bid. He establishes procedures for estimating quantities of materials and pricing the estimate for projects being bid. He supervises the estimators or "quantity takeoff" men. (These men work with scale drawings made by the architect or engineer and "take off" quantities of materials to be used in the project.) For effective bidding, he must be an expert on both the present and future expected costs of labor and materials. He also knows a great deal about your subcontractors and their capabilities and can award subcontracts accordingly. He should be aware of the costs from each project being constructed. He will then have the latest cost figures available to price the estimate for the job being bid. This is very important for bidding work. Knowing these costs makes you more confident in pricing the work being estimated.

2.3.3 Chief Project Manager

The chief project manager is responsible for all construction work. He supervises the individual project manager in charge of each job. He has to be a real fireball—knowledgeable, experienced, and an "hours mean nothing" type of person. He has to go where the work is, often miles from the home office. A great deal of travel is characteristic of the work to which the man holding this important position must adjust. He must establish procedures to keep all shop drawings, change orders, and other paper work under control, and if there is no general superintendent, he must help handle the job construction details.

2.3.4 Responsibility and Authority

The responsibility and authority of all the supervisory personnel in whatever position should be well delineated. All employees should know what authority these people have. When all people involved understand their responsibilities, confusion and instances when the ball may be dropped are eliminated. See paragraph 2.5.3.1 for more on this subject.

2.4 PROJECT MANAGING

The right project manager is one who has had successful experience in similar projects, a person who can accept the responsibility required for the type of project being constructed. The size of the project may or may not have a bearing on the capability required. The size, unless very large, and/or the complexity of the project are the governing factors. Some project managers can handle more details than others. They may have more experience and with this qualification can handle larger or more complex projects. Their temperament is also a very important factor. The reason for this is that some people they are working with, whether the owner, the architect, or anyone associated with the project, at times need a steadying influence, and the project manager could play that part. He must not lose control when things get tough. Project manager should be trained by starting on smaller and simpler jobs. They can handle more than one project. As they gain experience, they can move up to larger or more complex projects which help give them the experience to handle any job. As an example, a hospital is generally very complex and or specialized as compared to a warehouse, which can be very simple and does not generate many problems.

A project manager doing his first project that happens to be a warehouse will not have very many problems. On the other hand a project manager doing his first hospital should have some experience, perhaps as an assistant project manager on a previous hospital project. Hospitals, for instance, have several items installed in all of the patients' rooms, such as a toilet, gas, oxygen, and vacuum outlets; special electrical outlets; nurses' call systems, and so on. You will note these are all in one room. Other complex spaces include x-ray rooms, operating rooms, a CAT scan space, and other specialized areas. This is not the kind of project for a beginner. The coordination required is extensive, schedules have to be maintained and if you do not understand the details you will not understand the duration of scheduled activities. Some laboratories also fall into this class of complex work. A person new to this kind of project does not have time for on the job training, he must know what he is doing from day one.

A project manager also reviews all shop drawings (drawings prepared by the contractor, subcontractor, or supplier showing details of fabrication, erection, electrical or mechanical cuts) or sees to it that they are received on time for the contractor's review prior to submission to the architect for approval. It is well to work closely with the architect on shop drawing submission and approval procedures to ensure action in a timely fashion. The project manager prepares all change orders and works closely with the superintendent to keep the job running smoothly.

Purchasing of materials may be controlled directly by the project manager at the job site or handled through the main office. In either case, it is vital to have accurate records of quantities received at the site and installed in the work.

Quality control is a very important item that has to be kept in the proper perspective. Quality control means making sure that not only are your people doing the work in a workmanlike manner but the subcontractors are doing likewise. Quality control is a function of all the supervisory people connected with the project. The basic responsibility, in my opinion, rests with the superintendent. The project manager as well as the field engineers also share this responsibility. During the many tours through the project that any of them make they should look at the various items of work, going on to see to it that the work is being done properly. This means that they should be familiar with what is called for in the specifications and not let the job just float along on its own.

2.4.1 Project Start-Up

When it is known that you are the low bidder, your chief project manager should make preliminary plans to staff the project even though actual contract award may be weeks or months away. If you wait until the award, you may not have the time needed to move people around to staff the project at the last minute.

Selecting the right project manager for start-up will greatly affect the ultimate profitability of the contract. This may require taking a manager from another assignment even though it is not yet complete. Often the number two man welcomes the opportunity to take over in the final stages of project completion. Good project managers should be trained by coming up through the ranks, starting as an assistant field engineer, field engineer, then assistant superintendent. Some project managers can go through the superintending or even the estimating ranks. This type of program gives them a complete understanding of all facets of the business as well as an appreciation of what the other fellow is doing and how it affects what they are doing. It also might shape which phase of construction they really want to work at. Some people would rather work in the office and some would rather be in the field. This experience will give them an insight of which area they prefer to work in, thereby eliminating misfits on a job.

2.4.2 Award of Subcontracts

Awarding subcontracts, in our opinion, is a function best handled by the chief estimator. He is more fully conversant with the specifications, plans, and general scheme of the undertaking than any other person on the staff at this point. He can ensure that there will be no open ends or overlapping in subcontract awards. Some of the larger firms have their purchasing department handle the award of subcontracts. It is advantageous for the purchasing department to review with the estimating department to make sure they know what they must include or exclude from the subcontract. In the appendix is a subcontract form that can be used. This form will require you to fill in the date, the name of the subcontractor, the work covered, the project name, the Owner's name, list all addenda, the general contract date, the amount of money agreed to, and Article 14 which can be used for any additional provisions and special conditions required by the contract. This article can include Equal Employment Conditions, special government requirements, handicap requirements, and any other item required. The last page contains the signatures of the parties. As in all contract forms, your attorney and insurance carrier should review them before being used.

Price is not the only consideration in awarding subcontracts. Experience shows that you can afford to pay some firms more money because they adhere closely to the schedule you have set. You cannot afford to have your whole job delayed by a bargain subcontractor. For this reason, the chief project manager should review with the superintendents who have worked with the subcontractors being considered before the awards are made.

Some subcontractors may have to be bonded but others may not. This too has a bearing on determining which subcontractor has the lowest price.

It is important that when a project is finished the superintendent and the project manager rate the performance of the subcontractors on the project. This evaluation should be reviewed with the person who will award subcontracts on future projects so he can cull out those who do not perform properly. You can also evaluate what type and size projects the subcontractor can best handle. Some are good on simple projects and others on more complex ones or any type of job. These are important considerations in awarding subcontracts to ensure a smoothly running project.

2.4.3 Job In Progress

Coordinating the work crews, the subcontractors, and timing the delivery of materials are the responsibility of the superintendent. He does, however, work closely with the project manager in some of these matters. He schedules the work and sees to it that progress is maintained. He attends all meetings with the owner and the architect to resolve job problems.

2.4.4 Accountant

The accountant is a key to control and profitability. Your accounting system can be as simple or elaborate as desired, but it must provide enough information so that you know where the firm stands at all times. This can be a weak link. You cannot wait until a job is complete to find out you lost money. Corrective action has to be taken at the earliest possible moment. Various accounting and cost-control systems are detailed in Chapter 7.

2.4.5 Payment Requisitions

A schedule of values is required by the contract documents on most construction work. The payment requisition is based on this schedule of values. The contractor's requisition for payment is based on the amount of work put in place that month. Prompt submission of requisitions provides the payments to keep the business running. If you delay, payments will be late, and you may have to borrow money to keep going.

A business can be solvent but short of cash, and because of tight money situations, some firms have gone into receivership. By sending in requisitions no later than the day of the month specified in the contract, you generally get payment in time to take advantage of cash discounts. Many items carry a 2% discount. Concrete can have a $0.50 or $1.00 per cubic yard discount if paid within ten days, or whatever period of time the supplier specifies, after the billing date.

A typical project breakdown is shown in Fig. 2.4.5.1. Note the breakdown follows the Construction Specification Institute 16 Section format.

SCHEDULE OF VALUES

Item	Description	Scheduled Value
1000	General Conditions	$532,622
1507	Temporary Heat	$89,000
2050	Demolition	$10,500
2055	Remove Oil Tank	$4,226
2060	Interior Demolition	$8,500
2200	Excavation and Site Work	$690,718
2480	Landscaping	$76,416
2490	Landscape Existing	$3,000

Item	Description	Scheduled Value
3000	Concrete	$589,121
4000	Masonry	$1,009,325
5100	Structural Steel	$206,484
5200	Roof Framing	$210,686
5250	Wood Trusses	$39,982
5500	Miscellaneous Iron	$91,452
6100	Rough Carpentry	$36,708
6200	Millwork	$255,918
6210	Install Millwork	$208,232
6220	Millwork-Exterior	$30,000
6225	Wood Doors (in 6220)	$0
7100	Waterproofing/Dampproofing	$3,524
7150	Caulking	$13,800
7500	Roofing Skylight	$75,108
7555	Roofing Shingles	$188,274
8200	Hollow Metal	$131,593
8225	Hardware (in 8250)	$0
8250	Install HM Hardware	$50,966
8500	Windows, Glass and Glazing	$98,278
8550	Glass-Glazing	$15,756
8575	Greenhouse	$9,511
9250	Drywall (new)	$315,447
9270	Drywall-Existing	$32,952
9900	Paint and Glazed Wall	$95,260
9950	Paint Existing	$7,600
9500	Acoustic and Wood Ceiling	$116,482
9300	Ceramic Tile	$68,354
9700	Finish Flooring	$76,210
9750	Wood Flooring	$9,000
10100	Toilet Partitions	$5,800
10800	Toilet Accessories	$14,800
11400	Kitchen Equipment	$71,241
13000	Misc Spec-Curtain Track	$1,000
13025	Misc Spec-Tack & Chalkboard	$19,735
14200	Elevator and Dumbwaiter	$99,235
15400	Plumbing	$388,395
15500	Fire Protection	$291,361
15600	HVAC	$684,138
16100	Electrical	$818,874
18000	Fee	$386,326
	Total:	$8,181,910

FIGURE 2.4.5.1 Schedule of Values Format

Requisition for period of Apr. 10, 1990 to May 10, 1990

Item	Description	Schedule of Values	Previous Application	THIS APPLICATION Work put in Place	Material Stored	Work Completed and Stored	% Comp.	Balance Remaining
1000	GENERAL CONDITION	$532,622	$514,651	$17,616		$532,267	100%	$355
1507	TEMPORARY HEAT	$89,000	$89,000			$89,000	100%	$0
2050	DEMOLITION	$10,500	$10,500			$10,500	100%	$0
2055	REMOVE OIL TANK	$4,226	$4,226			$4,226	100%	$0
2060	INTERIOR DEMO	$8,500	$8,500			$8,500	100%	$0
2200	EXCAVATE & SITEWORK	$690,718	$688,208			$688,208	100%	$2,510
2480	LANDSCAPING	$76,416	$36,363	$24,364		$60,727	79%	$15,689
2490	LANDSCAPE EXIST	$3,000	$3,000			$3,000	100%	$0
3000	CONCRETE	$589,121	$589,121			$589,121	100%	$0
4000	MASONRY	$1,009,325	$1,009,325			$1,009,325	100%	$0
5100	STRUCTURAL STEEL	$206,484	$191,694			$191,694	93%	$14,790
5200	ROOF FRAMING	$210,686	$170,396	$40,290		$210,686	100%	$0
5250	WOOD TRUSS	$39,982	$39,982			$39,982	100%	$0
5500	MISC IRON	$91,452	$91,452			$91,452	100%	$0
6100	ROUGH CARPENTRY	$36,708	$35,000	$1,708		$36,708	100%	$0
6200	MILLWORK-INT	$255,918	$255,918			$255,918	100%	$0
6210	INSTAL MILL	$208,232	$145,000	$63,232		$208,232	100%	$0
6220	MILLWORK-EXT	$30,000	$30,000			$30,000	100%	$0
6225	WD DOORS (IN 6220)	$0	$0			$0		$0
7100	WTRPFG DMPRFG	$3,524	$3,524			$3,524	100%	$0
7150	CAULKING	$13,800	$13,800			$13,800	100%	$0
7500	ROOFING SKYLIGHT	$75,108	$75,108			$75,108	100%	$0
7555	ROOF SHINGLES	$188,274	$172,720	$15,554		$188,274	100%	$0
8200	HOLLOW METAL	$129,136	$129,136			$129,136	100%	$0
8200	HOLLOW METAL	$2,457	$2,457			$2,457	100%	$0
8225	HARDWARE (IN 8250)	$0	$0			$0		$0
8250	INSTL HM HARDWARE D	$50,966	$40,000	$7,500		$47,500	93%	$3,466
8500	WINDOW GLASS & GLAZE	$98,278	$96,729			$96,729	98%	$1,549
8550	GLASS-GLAZING	$15,756	$15,756			$15,756	100%	$0
8575	GREENHOUSE	$9,511	$9,511			$9,511	100%	$0
9250	DRYWALL	$315,447	$315,447			$315,447	100%	$0
9270	DRYWALL-EXISTING	$32,952	$26,476			$26,476	80%	$6,476
9900	PAINT & GLAZED WALL	$95,260	$94,206			$94,206	99%	$1,054
9950	PAINT & EXISTING	$7,600	$7,600			$7,600	100%	$0
9500	ACOUST & WOOD CEIL	$116,482	$116,482			$116,482	100%	$0
9300	CERAMIC TILE	$68,354	$68,072			$68,072	100%	$282
9700	FINISH FLOORING	$76,210	$76,210			$76,210	100%	$0
9750	WOOD FLOORING	$9,000	$9,000			$9,000	100%	$0
10100	TOILET PARTITION	$5,800	$5,800			$5,800	100%	$0
10800	TOILET ACCESSORIES	$14,800	$14,587	$213		$14,800	100%	$0
11400	KITCHEN EQUIPT.	$71,241	$71,241			$71,241	100%	$0
13000	MS SPC-CURTIN TRACK	$1,000	$1,000			$1,000	100%	$0
13025	MS SPC-TACK & CHALK	$19,735	$19,735			$19,735	100%	$0
14200	ELEV. & DUMBWAITER	$99,235	$99,235			$99,235	100%	$0
15400	PLUMBING	$388,395	$388,395			$388,395	100%	$0
15500	FIRE PROTECTION	$291,361	$287,447	$3,914		$291,361	100%	$0
15600	HVAC	$684,138	$681,403	$2,735		$684,138	100%	$0
16100	ELECTRICAL	$818,874	$818,874			$818,874	100%	$0
18000	FEE	$386,326	$386,326			$386,326	100%	$0
		$8,181,910	$7,958,613	$177,126	$0	$8,135,739		$46,171

FIGURE 2.4.5.2 Requisition Format

Also shown is a typical requisition form, Fig. 2.4.5.2. This shows how the amounts requisitioned for each item with previous payment, this payment requested, the balance remaining, and the amount retained. Notice that the requisition is first approved by the architect and then by the owner before payment can be made.

2.4.6 Project Wind-Up

The wind-up of a contract is the responsibility of the project manager. After substantial completion he sees that items on the punch list (usually small items of corrective or uncompleted work listed by the owner or architect during an inspection at the time of substantial completion), are done quickly. Prompt action on small items, even those not specifically part of the contract, will help you maintain a good relationship with both the architect and the owner. Retainages (money withheld for final completion and often paid after the owner has occupied the premises) should be released when due. The substantial completion date triggers the start of the warranty period; the sooner you finish the punch list, the fewer problems occur in warranty periods.

Some owners try to have the warranties start after final completion instead of substantial completion even though the specifications say substantial completion. Watch out for this condition.

In addition to the above, copies of all items required by the specifications to be turned over to the architect or the owner such as warrantees, guarantees, as-built drawings, mechanical and electrical manuals, test reports, and air balancing reports must be obtained from the suppliers or subcontractors and transmitted to the proper party as specified.

2.5 BUILDING AN ORGANIZATION

2.5.1 Multiple Responsibilities

Highlights of the duties of various people involved in a larger firm have been pointed out above. In smaller firms it is not unusual for one person to handle two or three of these functions.

When a firm starts out it may have only one estimator who takes off the quantities, prices the estimate, and may also do some of the purchasing. He may check shop drawings. The boss may do some of the subcontract awarding or also act as the project manager, visiting the jobs and performing the tasks of a project manager.

To be successful, one does not have to be topheavy with management people. Some small firms are one-man operations. No matter what the size of the firm, the basic work functions are the same. As in constructing a large building, the quantities are greater but the logistics are different from those in a small project. On bigger buildings items sometimes overlooked by growing firms are lower efficiency of production and the additional cost of getting materials to the areas where they are to be installed. Smaller jobs, having small crews and a working foreman, usually achieve lower costs. Larger projects, particularly those with tight schedules, involve bigger crews with less efficiency and nonworking foremen. Production is not always as great per worker. It takes more time and equipment to move materials and workers greater distances, both horizontally and vertically. These factors should be taken into account.

An alternative system of handling projects is sometimes used by smaller firms. The estimator who takes off the project also acts as its project manager. When he has only one project to handle, he can still have time to take off quantities for other projects. A disadvantage to this system occurs at bid closing time. The man must spend all his time with the new bid. This causes him to neglect the project he is managing, perhaps at a crucial point.

2.5.2 Hiring Experienced Personnel or Training People

In a small firm you expand by hiring assistants to help in estimating or supervising field work. Assistants should be hired on the basis of your assessment of their future potential as well as their present skill. This is not easy, and some mistakes will be made, but competent, dependable people are the key to future success.

Encourage self-improvement and training of your younger people through evening courses. Some companies pay part or all of the tuition cost for business-related courses that are successfully completed. You or someone in the firm must take the time to explain the interrelationship between what the individual is doing and how this relates to the overall workings of your organization. Simply explaining how to do the task assigned to him is not enough. When employees can understand the reasoning and relationship of what they are doing for the firm they are more eager to learn and they have a sense of being a part of the team. This gives them the drive to do better work.

Personnel of proven competence are essential for your success. People whom you have trained or worked with in the past are a known factor and their competence will govern to some extent the work you can handle. Expanding too rapidly with new or untried crews and supervisors may result in financial disaster.

The competence of personnel is so important it cannot be overstressed. Whether they be estimators, project managers, superintendent, field engineer, or other personnel, you must have people who produce good results.

Good and sufficient work forces are needed to complete a project whether they be trained through your own or other programs. In "Schedule A" from the A.G.C.A. are listed many training programs that can be very helpful.

If you are staffing a large job and are short of qualified people, it may be advantageous to break up good crews and spread the older employees among the new crew. Experienced company people can familiarize the new crew with your methods and quality of work.

As an example, you may have a project going that has an experienced project manager, superintendent, assistant superintendent, and engineers. If you start a new project and need a new crew, you can hire a superintendent and move the assistant superintendent to the new job to help the new people learn your systems and procedures. The same holds true for project managers as well.

2.5.3 Job Descriptions

All firms should have a job description for each position of authority. The following are some of the descriptions we have used. You can change them to fit your particular situation. You may consolidate some of them in a smaller firm or expand them in a larger firm. The key is to have a set made to suit your purposes. Job descriptions clearly establish the authority of each position and do not give someone an excuse to say "I did not know I was responsible for that."

2.5.3.1 Vice President for Marketing

Review leads on new projects from various sources:

a) Newspapers
b) *Dodge Reports* (to be furnished by Estimating)
c) *Engineering News Record*
d) Other periodicals that contain such information

Make calls to parties involved. Make calls on architects/engineers, owners and authorities trying to sell your construction services.

When we get involved with a project be responsible for:

a) Information the potential client wants
b) Budgets, with help from Estimating if required

Work with Estimating when necessary for preliminary estimates based on information developed from contact with owners.

Meet with owners, architects, engineers, consultants, and subcontractors when necessary to accomplish any of the above-mentioned goals.

On a regular basis, keep the president informed on the status of all projects you are working on and get direction from the president on some of the leads you develop.

2.5.3.2 Vice President in Charge of Estimating

Review construction reports for any potential jobs listed that may be of interest to bid. Review with the vice president in charge of marketing any potential projects he is working on and assist him with budgets and information needed.

Obtain plans and specifications for any jobs we will be bidding on.

a) Issue "Request for Bids to subcontractors."
b) Hire consultants or additional help needed for taking off quantities if necessary.
c) Be responsible for obtaining bid bonds for all projects being bid.

Prepare estimates and price same with the cooperation and help of the vice president in charge of project management and the general superintendent.

Oversee all the estimators in the department and coordinate their work for the quantity takeoffs they are working on.

Inform management staff of projects being bid and of the bid dates. Take subcontractor prices and put the bid package together as early as possible so that we do not have a last minute "crunch" at bid time.

Prepare information required for a data bank using parameter costs. All job information from work bid or completed in the past four or five years will be kept in the data bank. This information will give some accurate cost statistics to help budgeting and marketing for potential work and for work that is being bid.

Upon award of a contract to the company:

a) Award all subcontracts
b) Prepare subcontracts
c) Review same with project managers to ensure their completeness and understanding in order to avoid future problems.

Keep subcontractors list up-to-date.

On a monthly basis, keep the president informed on the status of all projects under his control being bid or considered to bid.

2.5.3.3 Chief Project Manager Be responsible for all project managers:

a) Assign project managers to the various projects to be constructed.
b) Systematize procedures of all projects. This will ensure all project records are in the same format. When a project manager is moved, he can easily follow the new project records.

Visit projects periodically and review with the general superintendent and the superintendents the status of subcontractors, as well as relations with the owners and architects.

Review the status of all projects with the general superintendent to ensure full cooperation on the operation of all projects.

Make sure all change orders and paper work of project managers is kept up to date.

Work with the estimating department, reviewing bid pricing and time scheduling on any project being bid.

Work with the vice president in charge of marketing to aid in reviewing scheduling, etc., on an as-needed basis.

Make sure the quarterly status reports are filled out on a timely basis.

On a monthly basis, keep the president informed on the status of all projects under his control.

Monitor flow of job costs data to accounting department for review and analysis.

Check that requisitions are being submitted on a timely basis.

Review any submittal process to ensure it is working well (e.g., shop drawings, schedule reviews, and requisitions).

Establish the routine to be followed for all project meetings, the standardized reporting format to be followed, and the time the meeting is to be held.

2.5.3.4 General Superintendent Be responsible for the construction aspects of all projects. Have a thorough knowledge of all facets of building construction methods necessary to make suggestions and to participate in discussions that result in the proper decision for the company and owner.

In conjunction with the vice president in charge of project management, establish the time schedule for each project.

Select the job superintendents. In conjunction with the superintendent, select engineers, assistant engineers, and the craft foremen to form a cohesive team so as to implement the planned schedule of time and costs for each job.

In conjunction with the job superintendent, determine construction methods and equipment to be used on the project. Visit the projects on a regular basis to make sure they are progressing on schedule and within the cost parameters. Review the construction methods being employed to see that the project is working out as originally planned or whether a different method would work better.

Review with the superintendent the equipment needed for the project. Work with your equipment manager to make sure we get the maximum use of our equipment before any outside equipment is rented or leased. Make recommendations to the president or vice president in charge of project management whether or not to buy or lease additional equipment.

Work with the vice president in charge of estimating and marketing to review the construction methods and time schedules on jobs we are bidding or negotiating.

Have current knowledge of work rules and an understanding of the collectively bargained agreements, if applicable.

Be responsible to the president. However, work very closely with the vice president in charge of project management, reviewing the status of the projects together to ensure full cooperation in the operation of all projects in this area.

2.5.3.5 Superintendent Report to the general superintendent.

Work with the project manager. The project manager gets the information needed by the superintendent to run his job properly. Be familiar with the contract with the owner so you know what the obligations are. Be familiar with the subcontractors' contracts so you know what their obligations are to us and our obligations are to them.

Be responsible for the construction phase of a project. This also includes supervising the field engineers, foremen, and others in our employ at the job site.

Develop with the project manager and general superintendent a construction schedule for the project.

Review all construction methods with the general superintendent well in advance of starting an operation to avoid future problems and delays.

Make sure the job is properly manned to keep the project on schedule.

Oversee and coordinate the subcontractors as well as making sure the job is properly manned by them to maintain the project schedule.

See that the materials needed by our forces and the subcontractors are received on time.

Inspect the work for contract compliance as to types of materials and quality workmanship. This includes our work as well as the subcontractors'.

Monitor costs to ensure they are within the estimate for both for quantity and labor units. Call any discrepancies to the project manager's attention so that there are no surprises at the tail end of a job.

Shop Drawings Review with the project managers which shop drawings are critical so they can get them prepared early on by the subcontractor in order to expedite and get them checked ahead of time so the material will be ready when needed.

Job Meetings Hold job meetings with the subcontractors to ensure everyone is pulling in the same direction. The frequency of these meetings should be governed by the size and complexity of the project. They should be held no less than every two weeks. The job meetings should not be confused with the project meetings, which are moderated by the project manager. The project meetings try to solve the problems that arise at the job meetings. The project meetings are attended generally by the owner, architect, engineers and the superintendent. These meetings should also be attended only by those subcontractors with major problems to solve. If all subcontractors attend the project manager's meeting then people not very involved may detract from the main problems to be solved.

Job Files Make sure they are updated on a constant basis, ensuring that the latest information and plans are on the job site.

Cooperate with the safety supervisor to help prevent accidents and also to the firm in compliance with state and federal laws.

Help train the engineers to be more adept at their jobs. Also, teach them to be good assistant superintendents so that they can advance to the position of superintendent when an opening occurs.

Make sure that relations with the owner, his representative, the architect and design engineers are kept on a high plane. Make sure you understand the instructions and answers you receive.

Schedule equipment, truck, and delivery requirements far enough ahead in order to prevent delays in construction or cause problems for other job needs from the yard.

Release equipment as soon as possible so other jobs do not have to hire additional equipment.

Reinforcing Steel Make sure all deliveries are checked. Our firm pays for erection on a ton-in-place basis and the steel delivered becomes the tonnage figures. Keep a running total on the job. In addition, keep track of mesh quantities.

When shoring or forms are rented, make sure they are counted when they come on the job and when they leave. This can prevent a lot of questions and improper charges when the material gets back to the supplier.

Other Items Keep a record of all items delivered to the job site such as:

Miscellaneous Iron Items

Door Frames

Hardware

Toilet Accessories

Extra Work Orders When you sign for subcontractor work, unless it is clearly for the firm's account or a Time & Material extra, it shall be noted as certifying for time only and not for payment.

At the end of each job, evaluate the subcontractors and turn in the list to the General Superintendent, using a scale of 1 to 10 with 5 as an average.

2.5.4 Student Cooperative Programs

Good experienced supervisory people are hard to find. In some areas of the country, colleges have a five-year cooperative system of education. Engineering and construction oriented class students attend school full time the first year. Equal periods of work and school make up the next four years. A period may be ten weeks in school and then ten weeks at work. During these work terms, a wise construction firm will utilize students' talents. Students learn more this way, because they relate what they are being taught in school to what they are doing in the business world and upon graduation they are experienced enough to assume responsible positions without a lot of additional training. This cooperative program provides an excellent source of reliable talent. These work periods give you an excellent opportunity to evaluate these students for future positions in your firm. Will they fit into your firm? Will they grow in stature with the firm? Are they willing workers? Are they ready to accept responsibility? These types of questions and others can get answered during the four year school–work period.

Students are also available during summer vacations and are another source of future help. The same evaluation that is used for cooperative students can be used for these summer employees. Our future needs are not only for hourly workers; we will also need additional good supervisory help.

Rotating your new people through various duties broadens them. You may be surprised to find a person hired for one phase is much more qualified and talented in another. Use your people where they can best utilize their strong points for it gives them a chance to show what they can do, and also makes them more valuable to the firm. People trained this way learn your system and are much more compatible with the rest of those in the firm. The firm always has a ready source of potential executives, and growth of the firm is ensured with capable back-up supervisors.

During this training period, the workers involved should be talked to in order to find out what aspect of the business they prefer. It is important for people to be happy with their work, and if they are used where they prefer to be, you have better employees who will work harder doing the type of work they prefer.

2.5.5 Managers

A contractor's success will sometimes come faster than his ability to train and develop his own people. He then must hire skilled people who have gained their experience with other firms in the industry. A good reputation makes it easier to attract first-rate people at the management level who have skill and a wide knowledge of people in the construction industry. A good manager is not going to leave a firm to go where his reputation will be damaged by some company not as good as the one presently employing him. This is true of skilled people in our classifications of work as well.

2.6 SAFETY AND EQUAL EMPLOYMENT

Safety is another area of importance to your firm. It is discussed in Chapter 11.

2.6.1 Job Description for a Safety Engineer

He is responsible to the President for all safety programs and actions of the firm. He will review the specifications for each project to note if there are any special safety requirements. He will establish the safety program for the project. He will review all safety requirements with the project manager and the superintendent.

He will carry out all safety inspections and will provide copies of his report to the President, project manager and superintendent. He will follow up to insure the corrective action is taken of all items noted on his report that are deficient. He will make sure that all accident reports are filed with the insurance carrier and with the proper authorities. He will review with the carriers our accident record to insure the rates charged by the carriers are proper. He will organize the job tool box safety meetings and insure they take place.

He will organize the company safety meetings including the safety award program and make sure the boss is available to attend and participate in these meetings.

In a smaller firm he would also have the duties of the Equal Employment Officer.

Equal Employment Opportunity covers minority hiring, minority business enterprises, hiring of women, women-owned businesses, hiring the handicapped, and other such categories. These are important items in our business. Many owners require goals of varying percentages which have to be striven for. There are also laws now promulgated by some municipalities that require a certain percentage of your work force to be residents of the municipality where the project is located. These require a lot of attention. In many cases it takes a full time person to keep up with the regulations and paper work and oversee the programs to make sure they are being implemented.

There are so many agencies involved and periodic changes that I will not try to enumerate them here. Suffice it to say, make sure you understand what is required so you can properly cover the item in your bid.

A book on this topic available in the Wiley series is; *Construction Contractor Survivor Guide by*: T. C. Schleifer.

2.7 CONCLUSION

Basic work functions are the same in the small firm and the large firm. Subordinates seldom see the overall picture in true perspective. Each person thinks his job is most important. This is good in some respects and makes him want to do his best. This is another reason to broaden all your supervisory people's experience so they understand their interdependency. The top man must see that no details are overlooked and all work is performed properly so that there is profit and growth at the end of each year. His subordinates must do the same with the people in their areas.

CHAPTER 3

Assessing Financial Capabilities

3.1 ESTABLISHING YOUR CAPITAL STRUCTURE

Establishing yourself as a contractor requires more than just a knowledge of construction work. Proper financing of your venture is essential or you may well be in trouble before you start. You must know how much capital is required, what cash needs will be, when your payments will come in, and alternatives that exist when things do not go according to plan.

Your capital position is vitally important for both short term and long term success. You should avoid tying up large amounts of money in all types of equipment. It is better to maintain a liquid position so you have the cash to pay bills on time and take advantage of discounts. This is money in your pocket. Buy only the smaller items of equipment, vehicles, and furniture that are needed for daily operations. Larger and more expensive purchases can be put off, by renting, until they impose no strain on your financial capability.

3.2 HOW TYPE OF WORK AFFECTS YOUR CAPITAL NEEDS

Capital needs are governed by the type and volume of work you plan to do. With annual contract volume in the $1 million or $2 million range, available capital should be one or two hundred thousand dollars. Type of work and number of items subcontracted can raise or lower this figure.

Simple buildings require less capital, particularly when much of the work is subcontracted. For example, a warehouse may have a minimal foundation, a concrete floor, and a structural steel frame with metal skin and roof. Your firm would do the foundation and the floor with your own forces. The steel frame, siding, and roof could be subcontracted. Your labor ratio would be low and cash needs would be lower. On the other hand, if the warehouse is complex and has a reinforced concrete frame and roof so that your forces would do most of the work, cash needs would be greater.

In addition to meeting payrolls, you would want to take advantage of discounts on reinforcing steel and concrete to stay competitive.

Buildings with extensive utilities and service requirements often have a higher percentage of labor and materials that would entail more capital. Some of the more complex buildings that require a lot of automated or specialized equipment may have lower labor ratios. Therefore each project must be reviewed on its own.

Subcontracting in most cases reduces cash needs on a complex project. However, this requires close supervision and scheduling to assure adequate quality control and timely completion. Knowing the best subcontractors is only part of the solution. A great deal of experience is required to coordinate the activities of subcontractors and prevent chaos at the job site and with your costs. See Chapter 8 for further details on supervising. See Figs. 3.2.1, 3.2.2, and 3.2.3 for cost and work allocation breakdown.

Code	Description	Item Cost	Resource
02200	Site Work and Excavation	$515,000	SUB
02610	Paving (in site work)	$0	
02250	Dewatering (Pump Only)	$12,000	GC
02710	Fencing	$10,100	SUB
02800	Landscaping Allow	$10,000	SUB
03300	Concrete	$880,515	GC
05100	Structural Steel and Deck	$804,670	SUB
05500	Miscellaneous Iron	$401,694	SUB
05700	Perimeter Walls/Railings(in 15)	$0	
07640	Metal Siding (Erected)	$36,186	SUB
06000	Rough/Finish Carpentry/Hardware	$2,041	GC
07500	Roofing	$6,575	SUB
09900	Painting	$195,780	SUB
15400	Plumbing	$100,000	SUB
15500	Sprinkler (In Plumbing)	$0	
16000	Electrical	$138,000	SUB
01000	General Conditions	$348,096	GC
	Subcontractor Bonds	$20,000	GC
	Sales Tax	$26,574	GC
	Total	*$3,507,231*	
	Work By General Contractor	$1,289,226	
	Work By Subcontractors	$2,218,005	

FIGURE 3.2.1 *Cash Flow Needs*

Code	Description	Cost	Resource
01000	General Conditions	$1,116,600	GC
02050	Demolition	$251,235	SUB
02200	Excavation and Site Work	$186,100	GC
02050	Asbestos/Contaminated Material. Allow	$46,525	SUB
02300	Piles/Caissons	$46,525	SUB
02800	Landscaping	$37,220	SUB
02500	Site Improvements	$74,440	SUB
03300	Concrete Foundations	$558,300	GC
03330	Concrete Structural	$977,025	GC
04210	Masonry	$3,070,650	GC
05120	Structural Steel	$418,725	SUB
05500	Miscellaneous Metals	$279,150	SUB
06100	Rough Carpentry	$55,830	GC
06200	Finish. Carpentry/Millwork	$186,100	GC
07500	Roofing	$279,150	SUB
07100	Waterproofing/Sealants	$186,100	SUB
07200	Building Insulation	$111,660	SUB
08150	Doors, Frames, Hardware	$139,575	GC
08500	Windows, Entrances, Glazing	$418,725	SUB
08360	Overhead Door	$93,050	SUB
09250	Drywall	$697,875	SUB
09310	Ceramic Tile	$158,185	SUB
09650	Relient Flooring, Carpeting	$214,015	SUB
09900	Painting	$148,880	SUB
10000	Miscellaneous Specialties	$148,880	SUB
11860	Rubbish Chute/Compactor	$27,915	SUB
12300	Appliances	$102,355	SUB
12350	Kitchen Cabinets, Vanities	$148,880	SUB
12500	Window Treatment	$27,915	SUB
11850	Parking Equipment	$9,305	SUB
14200	Elevators	$446,640	SUB
15400	Plumbing	$707,180	SUB
15500	Fire Protection	$669,960	SUB
15600	HVAC	$818,840	SUB
16100	Electrical	$1,330,615	SUB
	Total	*$14,190,125*	
	Work By General Contractor	$6,290,180	
	Work By Subcontractors	$7,899,945	

FIGURE 3.2.2 Cash Flow Needs

Code	Description	Item Cost	Resource
02200	Site Work and Excavation	$515,000	SUB
02610	Paving (In Site Work)	$0	
02250	Dewatering (Pump Only)	$12,000	SUB
02710	Fencing	$10,100	SUB
02800	Landscaping (Allow)	$10,000	SUB
03300	Concrete	$880,515	SUB
05100	Structural Steel and Deck	$804,670	SUB
05500	Miscellaneous Metals	$401,694	SUB
05700	Perimeter Walls/Railings (in 15)	$0	
07640	Metal Siding (Erected)	$36,186	SUB
06000	Rough/Finish Carpentry/Hardware	$2,041	SUB
07500	Roofing	$6,575	SUB
09900	Painting	$195,780	SUB
15400	Plumbing	$100,000	SUB
15500	Sprinkler (in Plumbing)	$0	
16000	Electrical	$138,000	SUB
01000	General Conditions	$348,096	GC
	Subcontractor Bonds	$20,000	GC
	Sales Tax	$26,574	GC
	Total	*$3,507,231*	
	Work By General Contractor	$348,096	
	Work By Subcontractors	$3,159,135	

FIGURE 3.2.3 *Cash Flow Needs*

On Figs. 3.2.1 and 3.2.2 you will note the items of work the general contractor does with his own forces and the work that will be done by subcontractors. Figure 3.2.1 shows an amount of $1,289,226 worth of work to be done with his forces. This amounts to 37% of the job cost. This is not an unusual percentage of work to be done by the general contractor. Figure 3.2.2 shows an amount of $4,290,180 worth of work to be done with his own forces. This amounts to 44% of the job cost, which is on the high end of the scale.

What does this mean? If you are doing this work you have to meet the payroll every week for the men involved. You may have to pay for material before the requisition comes due. You will have a retention of 10% at the outset which may reduce to 5% at some point. You can see by these numbers that the retention alone amounts to a sizable sum of money.

On the estimate in Fig. 3.2.1 the retainage alone could amount to $128,923. If you assume the labor cost is 50% of these items, you could need an additional amount of capital depending on the speed of the item.

On the Fig. 3.2.2 estimate, the retention could amount to $629,018 and the labor costs could amount to another sizable amount.

Note in Fig. 3.2.3 the amount of work to be done by the general contractor is negligible and his cost requirements are tied solely to the General Conditions item. The retention would be a maximum of $34,810, and the payroll amount would be negligible.

3.3 ESTABLISHING A LINE OF CREDIT

Establishing a line of credit is part of your financial picture. You can recognize the importance of discounting your bills to stay competitive and maintain a good rating. Delay in paying material suppliers can result in having material arrive at your job site C.O.D. Assurance in the form of assignments can be given to the supplier, but at this point you are in trouble.

Banks are helpful in the planning stage of setting up a business. They will review your goals and methods. Once you are ready to go, the bank will establish your credit limit for loans. You will know the sum that is in reserve to help strengthen your capital structure. Borrowing money for short periods to discount bills, even though the need is not urgent, establishes a desirable pay record. When your own cash gets tied up in delayed requisitions and slow receivables, your borrowing will be an established part of your business.

A periodic review of growth and financial status with your bank is good business. Open your books to them. Banks understand that firms doing well need money to operate. As you grow, your line of credit grows with you. If you are too guarded in revealing information, a banker will know this and be guarded in his appraisal of your firm. He may not extend your line of credit. His experience in evaluating and managing financial worth can be very beneficial to you.

3.4 YOUR BONDING COMPANY

Bonds are needed by a contractor when bidding work and after contracts are awarded. Selecting the right bonding company is a job for a thoroughly experienced insurance agent. Though it is nice to do business with a friend, be sure he is capable of putting together a practical and economical bond package. Otherwise, your growth capacity may be severely limited.

Detailed figures of assets and liabilities will be needed by the bonding company. In the beginning, your net worth governs bonding capacity. Later on experience adds to this capacity. This experience is based on many factors. First is rate of financial growth. Second is general reputation. Are projects completed on time? Is there frequent litigation? Are bids consistent, neither too low nor too high? The bonding company will learn a great deal about your firm and it is imperative to be completely honest. After all, they are going on the line for you.

3.5 NEED FOR OUTSIDE ACCOUNTANTS

Bonding companies, banks, and other outside organizations require audited figures. This necessitates employing outside auditors or certified public accountants (C.P.A.s).

In-house accountants are essential for proper control of your daily bookkeeping, cost control records, monthly or quarterly statements and other financial data. However, their close day to day familiarity does not always give objective results. Experienced c.p.a.s can recommend improved accounting methods to keep pace with the growth of your firm as well as keep you informed on new rules and regulations that come into being on occasion. They are also very beneficial in planning for tax matters as well as helping you file returns or filing them for you. In the event the Internal Revenue Service questions your return information, they are very helpful in this area and will generally review the problem with the IRS for you many times. Their advice can save you a lot of grief and additional tax payment because of their understanding of the tax laws.

Successful financial management always requires the use of outside accountants. Your accounting firm should assist you in the preparation of your statements and projections to show the bank and your bonding company. It is better when an outside c.p.a. is used because the statements can be certified, which give more credence. The following is a typical cover letter from a statement:

The Board of Directors
XYZ Construction Company, Inc.:

We have examined the balance sheet of the XYZ Construction Company, Inc., as of December 31, 1989, and the related statements of operations and retained earnings and changes in financial position for the year then ended. Our examination was made in accordance with generally accepted auditing standards, and accordingly included such tests of the accounting records and such other auditing procedures as we considered necessary in the circumstances.

In our opinion, the aforementioned financial statements present fairly the financial position of the XYZ Construction Company, Inc., at December 31, 1989, and the results of its operations and the changes in its financial position for the year then ended, in conformity with generally accepted accounting principles applied on a basis consistent with that of the preceding year after giving retroactive effect to the change, with which we concur, in the method of accounting for recognition of joint venture earnings as described in note 11 to the financial statements.

3.6 CASH FLOW

Cash flow is very important to your operation. Costs of interest on borrowing money can have a great affect on your earnings; hence it is important to keep your borrowing to a minimum. You should project cash flow by listing the receivables expected for the next six months. You then project what your expenses will be for the same period and then compare the results to see whether additional cash is or is not needed during this period.

Substantial funds are generally tied up in retainage on all contracts in amounts from 5% to 10% of the contract amount. Payrolls have to be met weekly as well as paying the taxes and fringes involved with the payroll. Material bills, depending on the terms, must also be paid periodically and also cause a depletion of cash.

It is important that your monthly requisitions be submitted in a timely fashion so that payment will be received promptly. You must also check to make sure these requisition payments are received per the contract. Today most contracts provide for interest to be paid on late payments. On substantial and final completion when final payments are due, you must make sure that last minute items are taken care of. Sometimes a small item can hold up a large payment. When you weigh the loss of the use of the money involved, it can be seen that the item should be taken care of immediately to get the payment released.

It is important to keep a check on your requisition payments. Some owners are lax in meeting their payment obligations and have to be constantly reminded that the payment is due. Before you sign a contract, you should know where the funds are coming from to ensure they are available when due.

The A.I.A. in their latest A201 General Conditions have included some new language in this regard that reads as follows:

> The Owner shall, at the request of the Contractor, prior to execution of the Agreement and promptly from time to time thereafter, furnish to the Contractor reasonable evidence that financial arrangements have been made to fulfill the Owner's obligations under the Contract. [Note: Unless such reasonable evidence were furnished on request prior to the execution of the Agreement, the prospective contractor would not be required to execute the Agreement or to commence the Work].

A book in the Wiley Series that refers to where funding comes from is *Construction Funding: Where the Money Comes From,* Second Edition, by C. A. Collier and D. A. Halperin.

3.7 ASSESSING AND MAINTAINING FINANCIAL CAPABILITIES

Capital is increased either by investing more money in the business or by retained earnings. Therefore, making a profit helps retained earnings. Excessive salaries and bonuses should not be paid out because this causes a drain on the finances of the business. Overhead is one area that has to be watched at all times. It creeps up and up to dangerous levels unless proper controls are exercised.

Some contractors take equity positions on projects they build. This can be dangerous if not properly thought out. First, even though you do not get cash, you still have to pay a tax on the amount earned. Second, if your after tax dollars is what you leave in the project, you have not provided money to take care of your overhead. This puts a drain on your existing capital and reduces your cash position.

3.8 JOINT VENTURES

Joint ventures can be beneficial, particularly on a large project. Cash is needed to finance this kind of venture however. Generally an amount of start up capital required for the project is determined by the partners and each participant puts up cash in the same percentage as his share in the joint venture. Joint ventures are discussed further in Chapter 13.

3.9 CONCLUSION

Learning the ins and outs of the financial world is essential to the successful operation of a small, medium, or large business. Rapid growth requires a close working relationship with your bank, your bonding company, and your outside accounting firm.

CHAPTER 4

Estimating

4.1 BIDDING FAMILIAR WORK

You are more competitive when bidding types of work with which you have had experience. Pitfalls can be avoided if you have solved the same problems before. You know the costs and the latest construction techniques for speed and economy. Working for the same owner gives the additional advantage of knowing what he expects. You can give him good service even when contracts include different types of construction, and you know you will be paid.

4.1.2 Bidding Unfamiliar Work

Owners and architects whom you have satisfied in the past may request that you bid on unfamiliar types of work. If not extremely busy, you generally accept the challenge. The project might be quite unrelated to your experience: an underground structure, a nuclear laboratory, some phase of an atomic power plant, a missile facility, a high rise project with architectural concrete, prestressed concrete, or any of many other special types.

4.2 MULTIPLE CONTRACT PROJECTS

If the opportunity to bid on a multiple contract project comes up, you must take into account the difficulties of controlling such work. Most construction work is bid on a conventional single contract basis. A general contractor has full responsibility. He selects subcontractors and has direct managerial and financial control over them. Under the multiple contract system, the owner or architect takes separate bids from electrical, mechanical, and other specialty contractors and awards contracts on a direct basis. You now have three or more prime contractors on the job site, none of whom have any contractual relationship with each other. It is not unusual for two or more contractors to go about their merry way, setting their own pace, and the devil may care about coordination or time. Yet the entire project must go ahead and be completed on schedule. In desperation, the general contractor for the structure will try to coordinate all work schedules, but he has no direct authority or monetary control. When bidding multiple contract projects with unknown specialty contractors, allow a contingency for probable extra costs. Talk to the architect or owner before bidding to see if any provision for coordination and direction has been established.

A somewhat better situation exists where the owner takes separate bids and then assigns them to you. These contractors now become your subcontractors. Keep in mind that you are now doing a larger management job, and your profit should be figured as if you were bidding the complete project in the conventional way. Most owners specify and pay an additional fee, usually stipulated in your bid, for handling these assigned subs.

4.3 SELECTIVE BIDDING

As stated earlier, you are more competitive when bidding on familiar work. A factor to be considered when selecting which project to bid on is the owner. If you have had previous experience with him and had a good relationship on the project, this definitely would be a factor. The reputation of the owners also should be considered in this process. You are contracting to build a project, not cause a lot of trouble which could even lead to law suits. The people you have available and their experience is also a factor to consider.

The litigious society that exists today must be very strongly considered when working for some clients. You must know who they are and what their reputation is before bidding on work for them.

4.3.1 Nonlocal Projects

On projects outside your local area, the chief estimator must gather all the pertinent information necessary to estimate the costs. A useful list developed a few years ago by the A.G.C.A. is shown in Fig. 4.3.1. This list includes space for local subcontractors, material suppliers, distance to the nearest rail siding, and availability of water, sewer, electric power, and gas. In addition, the list will remind you to check on wage rates, duration of union contracts, if applicable, and locations of union headquarters and building trades councils. The local chapters of the A.G.C.A. are a valuable source of information on any questions you may have in gathering facts you need. *Engineering News Record* periodically publishes wage and material rates for various cities in the country. This, too, can be very helpful. Computer services are also becoming available for this type of information.

OUTLINE FOR REPORTS OF SITE INVESTIGATIONS

HEAVY & UTILITIES CONSTRUCTION

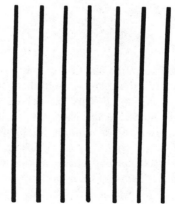

FIGURE 4.3.1 *Site Investigation Format*

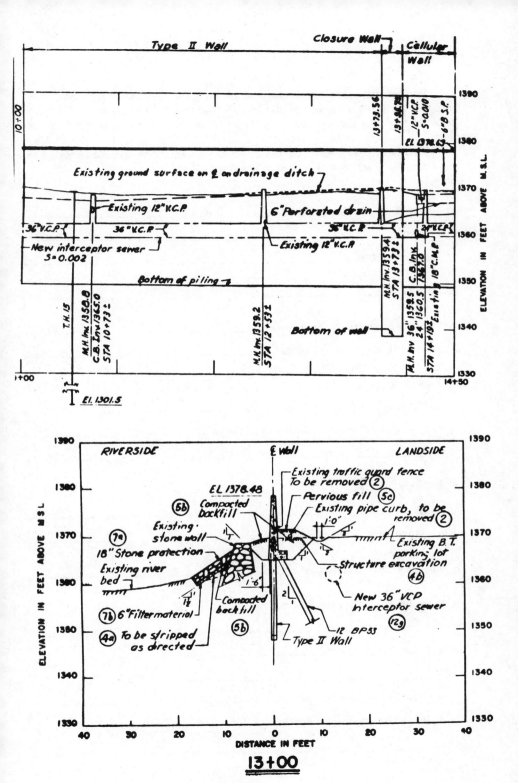

FIGURE 4.3.1 (cont.)

Site investigations are essential in planning construction operations. The following outline for reports on site investigations is recommended as a planning guide starting with the first review of the work site to beginning the mobilization phase.

This outline has been planned as a work book. Complete the checklist in this book for a permanent record in the project file.

1. Name of project _____

   Owner _____

   Location of project _____

2. Date of visit _____

3. Distance to closest towns or cities _____

   Sizes of closest towns or cities _____

4. Highways:

   a. Type and surface condition _____

   b. Capacities of bridges or load restrictions _____

      _____

5. Railroad facilities:

   a. At site _____

   b. At railhead _____

6. Nearest commercial airport _____

7. Barge lines and river conditions:

   a. C.F.S. flow in river _____

   b. River forecast office _____

      _____

   c. Scour conditions in river _____

   d. Barge traffic; channel depth; lock dimensions _____

      _____

FIGURE 4.3.1 (cont.)

    e. Commercial carriers (if any) _____

       _____

    f. Barge season _____

    g. Flood hazard _____

8. Haul roads, distance and condition:

    a. From railhead _____

    b. From barge dock facility _____

    c. From various material sources _____

       _____

       _____

9. Power connections:

    a. Name and address of supplier _____

       _____

    b. Closest installation and capacity description _____

       _____

    c. Cost of extension or installation on job _____

    d. Secure copy of rate schedule   Check when completed ☐

10. Telephone communications:

    a. Name and address of supplier _____

       _____

11. Land rental, ownership and availability if Owner does not supply adequate working area:

    a. At railhead _____

       _____

    b. At project _____

       _____

12. Determine extreme weather and length of working season _____

   _____

FIGURE 4.3.1 (cont.)

13. Contact local AGC:

    List names _____

    Addresses _____

    _____

    Telephone numbers _____

    a. Question for any unusual working conditions

    Check when completed ☐

    b. Unions and hiring hall locations _____

    _____

    _____

    c. Health, welfare, pension, vacation funds _____

    _____

    d. Travel pay and dispatch points _____

    _____

    e. Secure copies of current contracts with pay schedules

    Check when completed ☐

    f. Supply of common and skilled labor in area _____

    _____

    _____

    g. Predominant industry; agriculture, timber, mfg., mining, etc. _____

    _____

14. Housing: availability; adequacy; location with reference to job site _____

    _____

    _____

    _____

15. Trailer facilities: location; adequacy; availability _____

    _____

    _____

FIGURE 4.3.1 (cont.)

16. **Local subcontractors or suppliers:**

   a. **Adequacy of plant; method of delivery; reputation**

   Aggregate _____.

   _____

   Rock and riprap materials _____

   _____

   Clearing _____

   _____

   Painting _____

   _____

   Mechanical _____

   _____

   Electrical _____

   _____

   Ready-mix plants _____

   _____

   Rental haul trucks _____

   _____

   Grouting _____

   _____

   Seeding and sodding _____

   _____

   Bituminous surfacing and roadwork _____

   _____

   Well drillers _____

   _____

   b. **Secure copy of local telephone directory**   **Check when completed** ☐

FIGURE 4.3.1 (cont.)

17. **Visit owner's office:**

    a. **List names and titles; telephone numbers of persons contacted**

    Name _____

    Title _____ Phone _____

    Name _____

    Title _____ Phone _____

    Name _____

    Title _____ Phone _____

    Name _____

    Title _____ Phone _____

    Name _____

    Title _____ Phone _____

    b. **Discuss general job requirements**   Check when completed ☐

    c. **Clarify questioned items in specifications**   Check when completed ☐

    d. **Discuss specifics: stream pollution regulations, safety requirements, construction easements, landmarks, as needed**   Check when completed ☐

    e. **Examine special reports, etc. Secure copy if possible; otherwise make extracts of pertinent features**   Check when completed ☐

    f. **Secure pictures from owner as available**   Check when completed ☐

    g. **Request conducted tour of proposed work**   Check when completed ☐

    h. **Secure weather and streamflow information**   Check when completed ☐

    i. **Determine acceptance of streamflow for mix and cure purposes**   Check when completed ☐

    j. **Examine all cores and logs of test holes available**   Check when completed ☐

    k. **General discussion of geology**   Check when completed ☐

FIGURE 4.3.1 (cont.)

18. Describe site: Take photographs of all borrow areas; plant area; general construction areas; extent of clearing and grubbing required

_____

_____

_____

_____

_____

_____

_____

19. Determine availability of potable water

   a. Secure approximate costs of drilling and casing well _____

   _____

   _____

20. Work camp (if required):

   a. Recommend location _____

   b. Nearest utilities to area _____

   c. Distance from worksite _____

21. Recommend location of contractor's worksite:

   a. Locate office, shops and sheds _____

   _____

   b. Estimate work required to prepare site _____

   _____

   c. Locate batch plant and aggregate area _____

   _____ _____

FIGURE 4.3.1 (cont.)

(1) Estimate site preparation _____

_____

(2) Determine foundation conditions _____

_____

d. Locate storage areas for materials _____

_____

e. Recommend temporary roads _____

_____

f. Is local road surfacing material pit available   Yes ☐   No ☐

22. Investigate ground-water conditions, natural drainage areas and features which may be revised or relocated to assist in maintaining dewatered operation

_____

_____

23. Recommend best location and type of cofferdam to fit the job situation

_____

_____

24. Determine if terrain and working conditions will allow or require specialized equipment

_____

_____

FIGURE 4.3.1 (cont.)

NOTES AND COMMENTS

Compiled and published by the Associated General Contractors of America, 1957 E Street, N. W., Washington, D. C. 20006

FIGURE 4.3.1 (cont.)

4.4 ADVANTAGES OF A CONSTANT BACKLOG

A steady flow of work is the goal of every contractor. Peaks and valleys reduce profit. However, avoid the temptation to take work too cheaply (at a low profit margin) just to keep a crew busy during a slow cycle. Price work as it should be priced for adequate profit.

Two things tend to happen to your estimates when you need work. First, unit prices are shaved to a minimum because you want to be the low bidder. Second, the percent profit figure is reduced. This creates a bad situation. Unit costs are now difficult to meet, and losses may occur on some items of work. The profit factor has been reduced to the point where there is nothing for contingencies. You wind up with no profit or a loss on the job.

At first glance this may not seem too bad because you kept your crew busy. But in reality, the "break-even" job was a losing one, when you consider your own time and the cost of operating the business. The cost of operating the business is covered by profits, and if there are no profits you have a losing year.

Once you have a reasonable backlog of work, don't stop bidding. Keep estimating and put on a fair profit. The jobs you get will produce a good return, and the ones you lose will provide useful information to your estimating staff.

On examining his successful bids, a contractor will find that in 90% of the cases, he could have added 1% or 2% more profit and still have been the low bidder! Panic in bidding costs money. This is another reason to try to keep a backlog by steady bidding for new work.

4.5 THE ESTIMATE

When it is decided to bid on a particular project, plans and specifications are procured. The chief estimator must establish a system of bid preparation that can be adapted to any type of work and can be carried out by various subordinates. The invitation for bids should be carefully read to determine what is expected from you on bid day. Items such as number of copies of the bid to be submitted, whether a bid bond or other type of security is required, and finally the time and place of the submittal should be noted. You also learn whether or not the bids will be publicly opened or what other arrangement will take place. The bid form also may have one bid item or require various breakdowns or unit prices. These requirements will determine whether or not your bid takeoff must be differed to meet this condition. Unit prices and alternates also affect the way your takeoff is broken down for pricing purposes to meet the bid requirements.

There are other books in the Wiley series that are available that refer to bidding and estimating; listed are three of these books:

Construction Bidding For Profit by William R. Park
Construction Estimating For General Contractors by Leo Diamond
Construction Cost Estimates by Leo Diamond and C. R. Tumblin

4.5.1 Specifications

The next item to be studied is the specifications. Notes should be made, as they are read, on what is required for job personnel and insurance requirements. What municipal, county, state, or federal rules or regulations may apply. Who is responsible for what insofar as inspection and testing needs. Which temporary facilities are to be furnished by you or others and whether or not they are available. The aforementioned items are a sample of what to look for and note. There are others, of course. These are normally in the general or special conditions section of the specifications. The form of contract you will be expected to sign, if you are the successful bidder, is usually also included in this part of the specifications.

4.5.2 Special Conditions

Study any pages or paragraphs of "special conditions" carefully. Sometimes these are so nebulous or restrictive that you become responsible for functions beyond the scope of construction. You may become involved in the success or efficient use of the structure.

Such items as hold harmless clauses, which include coverage even if the other party is negligent, become dangerous. Continuing work even when payments are delinquent is another.

After the general conditions, the technical sections of the specifications follow. Notes should also be made when reading these sections on unusual items that may be required. A subcontract list is also prepared as you go through the various sections while noting whether a breakdown of some of the sections is needed to take care of cases where multiple items are put in one section. In some areas, for instance, mechanical subcontractors will do their own excavation and concrete work, required for their section, whereas in others they may not. Your experience helps you here. You may have to take off these items and price them so they are included in your bid. This condition may also be true in other sections of the specification.

If your firm does the concrete or masonry work, you must note the types of brick, mortar, forms, concrete types, and strengths required as well as other materials so you can start getting material prices to help complete your estimate on time.

This type of review helps you to understand many of the requirements of the contract as well as knowing more about the type of workmanship required. Items such as the type of brick bond which affects the cost of laying the brick, or the types of concrete finishing and rubbing are important to know because they affect your labor costs to a great degree.

Plan to do a lot of research and ask yourself many questions when preparing an estimate. Typical questions can include:

1. Is sheathing or shoring required?
2. Are unusual foundations required?
3. Are slabs so heavy that they require special shoring for form work?
4. Are the reuses of forms minimal?
5. Are the tolerances normal or tight?
6. Is special equipment required?
7. Do structural openings have to be left for the installation of this special equipment?
8. Does post tensioning require alternate bays to be poured?
9. Are special materials needed?
10. Is brickwork in long runs or narrow piers?
11. Is special aggregate specified for the concrete?
12. What is the heaviest load to be lifted into place?
13. Does concrete placement require pumps, power buggies, or a crane and bucket?

These are some of the important questions that you must evaluate that can affect the project cost and time schedule. There are, of course, others that have to be considered depending on the project. Allot plenty of time for these complexities as well as reviewing the time required for the delivery of these special items.

Don't underprice specialized work or you may win an unprofitable job, thereby weakening an otherwise healthy firm.

4.5.3 Escalation

A complex project sometimes will take two, three or more years to complete. On these projects contingencies must be added to cover such items as labor increases and possible price escalation for materials. The project schedule must be studied, and the labor required for each contract period evaluated so as to know the percent of labor carried for each of these periods. Once this has been established then labor escalation can be added in a calculated manner rather than a hit-or-miss guess.

The same procedure should be utilized for materials required to come up with proper contingencies for material escalation. See Fig. 4.5.3 for an example of how to calculate escalation.

ADDITIONAL COSTS FOR ESCALATION AFTER FIRST YEAR

Item	Work Remaining Starting	Number of Units Remaining	Added Cost Per Item Or Unit	Escalated Costs
Concrete Material	2nd Year	12,000 CY	$1.00/CY	$12,000
Concrete Material	3rd Year	5000 CY	$1.25	$6,250
Concrete Labor	2nd Year	$100,000	7%	$7,000
Concrete Labor	3rd Year	$45,000	5%	$2,250
Reinforcing Steel Material	2nd Year	900 Tons	$40/Ton	$36,000
Reinforcing Steel Material	3rd Year	375 Tons	$55/Ton	$20,625
Reinforcing Steel Erection	2nd Year	900 Tons	$30/Ton	$27,000
Reinforcing Steel Erection	3rd Year	375 Tons	$35/Ton	$13,125
Structural Steel Material	2nd Year	700 Tons	$100 Ton	$70,000
Structural Steel Material	3rd Year	100 Tons	$100 Ton	$10,000
Structural Steel Erection	2nd Year	700 Tons	$50 Ton	$35,000
Structural Steel Erection	3rd Year	100 Tons	$50 Ton	$5,000
Total Escalation Costs To Be Added				$244,250

FIGURE 4.5.3 *Escalation Format*

4.5.4 Subcontractors and Material Suppliers Lists

The subcontract lists that you have prepared as you read the specifications is now referred to. You should have a list of subcontractors and material suppliers classified by the materials they supply or the trade work they perform. Each material supplier or subcontractor should be numbered under the classification in which they are found. This helps expedite making up the list of people you will send notices to. For instance, under plumbing you may have:

1. Ace Plumbing Co.
2. Bravo Piping Co.
3. Colony Mechanical Contractors

etc., with their addresses and zip codes noted. When you fill out the sublist under plumbing, you merely show 1 and 3, which means a secretary will send a card to Ace Plumbing Co and Colony Mechanical Contractors notifying them you are bidding on the job and requesting them to submit a price to you. It also states when the price is due and who the owner is. Some firms use a two-part card requesting a reply so they know if the subcontractors are or are not going to bid on the project. This system saves writing the name of all the subcontractor invited beside each trade on your sublist and helps you know whom to expect prices from on bid day. You can also determine which trades sections require a spread sheet to be prepared to evaluate the bids properly.

4.6 QUANTITY TAKEOFF

You are now ready to take off the quantities for the work you expect to perform with your own forces or takeoff areas and quantities to make up estimates as checks against various subcontract prices you may get or estimate a cost for those you may not get.

A uniform system of taking off and recapping quantities is essential. This uniformity helps in the preparation of the cost breakdown for payment and cost-keeping purposes.

There are many ways to take off the quantities of a project. They should generally be taken off in the same sequence in which the project is carried out. This helps eliminate the possibility of error or omission. For example, concrete footings are placed first. Footings are followed by walls. Columns and slabs are next. Exterior walls are followed by interior partitions. My personal preference, which is the exception, is to takeoff the foundation quantities first. During this takeoff you become familiar with the bottom elevation of footings, piers pits, etc., as well as the elevation at grade. This helps your estimators become more familiar with what the cuts and fills will be, the slope of your banks or whether sheathing may be required. When you combine the information of these two items you also know how you can pour the foundation walls—whether the concrete can be chuted in from the truck or has to be pumped or placed with a crane. Access to the hole can also be more easily determined. Therefore the second item of takeoff is the excavation.

The superstructure follows and your notes will again help in how this takeoff is made because in the case of a concrete frame you may have different items to consider such as inserts, different strengths of concrete for the columns or beams, ramps and floors. Sometimes a special color of concrete is used in a particular area and this too is important to know.

In the case of a structural steel frame you may have fireproofing to consider. Special protection of materials and of the public are needed for some of the spray fireproofing materials used. Make sure this item is covered. The slabs may be on steel deck with concrete fill or could be formed concrete in each bay. It is the sequence as well as the material that is important.

The next item to be taken off is the skin of the building. The takeoff here will vary greatly with the type of skin. It is important, however, to note that in the case of brick, the type of bond, size of piers, area of panels, and the type of lintels and sills can vary your labor costs tremendously.

Once the skin has been quantified, the interior partitions and finishes are tackled. Again, the finishes must be noted to reflect accurately the costs you must put into the estimate to do the work properly.

This grouping also assists in setting up the cost breakdowns for cost-keeping record and also in the estimation of insurance costs to be added to the bid by utilizing the grouping to keep similar items within the insurance classifications by group.

Once the quantities of work that you do with your own forces are tabulated, they are then assembled on what we call white sheets, Fig. 4.6.1. These are fairly standard and the one illustrated here lists the item, unit of measure, quantity, unit price for material, cost of material, unit price for labor, and cost for labor. The sheets, as a rule, are kept numbered and an attempt is made to keep all concrete in one group of sheets, excavation in another, for example. This allows you to add up the numbers quickly for any particular item if you want to make a comparison against a sub-bid you may receive for one of these classifications. When the quantity estimate is completed, it is priced by the chief estimator, alone or in conjunction with the boss and/or the chief project manager or in some cases with a superintendent. Prices are based on past experience, previous costs and best judgment. New or unusual items must be priced with great care. Take time to break down the operation into steps and then price each step. You can come up with a close actual estimate this way. Experience shows that the unusual items are generally priced properly. It is the "run of the mill" items where large quantities are involved that require caution. A few pennies per unit can amount to a lot of money. These items are, as mentioned earlier, grouped together in categories such as excavation, concrete, masonry, carpentry, and finishes. Upon completion of these sheets, they are transferred to the recap sheet (See paragraph 4.9.1).

Concurrently, a time schedule is prepared and the overhead sheet is started. The time schedule determines the length of time supervisory people are needed on the job. An office setup may be required, and the length of time this setup is needed is established as well as how long equipment is required for the project. Other items such as temporary power, temporary heat, and water needs are determined by the use of the schedule. General conditions vary with the type of work, time breakdowns and other items. See Fig. 4.6.2. By standardizing and listing usual items you rarely forget an item. Special items noted from the general and special conditions can be filled in using the blank spaces available.

FIGURE 4.6.1 Estimate Sheet

4.7 IN-HOUSE QUANTITY TAKEOFF

Quantity takeoff necessitates close study of every detail in the plans and specifications. By having your own estimators do this work, they can uncover construction problems that might otherwise pass unnoticed because they know your system.

The estimate can be priced with more precision. People in your office will be more familiar with the project and better able to answer last-minute questions when the bid is being assembled.

The final advantage of doing your own quantity takeoffs is that the items will match perfectly with your own cost-keeping system. This makes it much easier when breaking down the bid to set up the cost system and requisition breakdown.

A.

THE VOLPE CONSTRUCTION CO., Inc.

	EXT.	ADD.	EST.
1			
2			

ESTIMATE

Job _____ Bids Close _____

_____ Archt. _____ Cu. Ft. _____ Cost per ft. $_____

Branch GENERAL CONDITIONS/OVERHEAD	Units	Quantity	Unit Price Materials	Cost Materials	Unit Price Labor	Cost Labor
Project Manager						
General Super. (part time)						
PROJECT Superintendent						
" " Assistant						
" Engineers						
" " Assistant						
" " Specialty						
" Accountant						
" Timekeeper						
" Clerks						
" Watchmen						
PERMITS/FEES - Building						
" " Sidewalks						
" " Street						
Registered Survey Crew						
TEMPORARY Field Office						
" COW Office						
" Telephones #						
" Lockers - Lab.						
" " - Carp.						
" " - Other						
" Storage Sheds						
" Heat all above						
" Toilets/Clean #						
." Water						
" Light/Power						
" Electric Service						
" Power for Steel						
" Heat/Owner's Plant						
" Stairs/Ladders						
" Staging for Subs						
" First Aid Station/Nurse						

IP-5,000 - 7-70

FIGURE 4.6.2 *General Conditions Sheets*

4.7.1 Outside Quantity Takeoff Service

A quantity takeoff can be purchased from firms specializing in this service. Properly understood and used, this data can be quite useful. It can save time and if you desire will provide a check on your own estimator's quantities.

Purchasing these quantities has the disadvantage that you may establish prices without the full understanding and feeling for the project that comes with hours of intensive study during the takeoff process.

THE VOLPE CONSTRUCTION CO., Inc.

	EXT.	ADD.	EST.
1			
2			

B.

ESTIMATE

Job_____ Bids Close_____

_____ Archt. _____ Cu. Ft. _____ Cost per ft. $_____

Branch GENERAL CONDITIONS/OVERHEAD	Units	Quantity	Unit Price Materials	Cost Materials	Unit Price Labor	Cost Labor
WINTER CONDITIONS -						
Enclosures - Masonry						
" Concrete						
" Openings						
Heat/Portable Units						
Snow Removal - Building						
" " Site						
Enclose for Spray Fireproofing						
Project Signs						
CPM/Updating						
Field Office Supplies						
Engineering Equipment						
Project Drawings						
Shop Drawings						
As-Built Drawings						
Trucking						
Periodic Clean-Up						
Trash Container/Dispose						
Final Clean-Up						
Clean Glass/Aluminum						
Replace Broken Glass						
Project Photographs						
Protect Finish Floors						
Rubbish Chutes						
SAFETY- General OSHA						
" Security Fence						
" Guard Rails-Perim.						
" " " Forms						
" Fire Protection						
INSURANCE - Design E & O						
INSURANCE - Builders Risk						
" F/T/V NOC						
" Hold Harmless						
" Other						

IP-5,000 - 7-70

FIGURE 4.6.2 (continued on next page)

THE VOLPE CONSTRUCTION CO., Inc.

	EXT.	ADD.	EST.
1			
2			

C.

ESTIMATE

Job_____ Bids Close_____

_____ Archt. _____ Cu. Ft. _____ Cost per ft. $_____

Branch GENERAL CONDITIONS/OVERHEAD	Units	Quantity	Unit Price Materials	Cost Materials	Unit Price Labor	Cost Labor
TESTING/INSPECTION						
Concrete - Field						
" Plant						
Earthwork						
Reinforcing Steel						
Piles						
Structural Steel - Field						
' " Plant						
Metal Deck						
Shear Connectors						
Precast Plant						
SAMPLES - Masonry						
" Concrete						
Small Tools						
Boats, Coats, Helmets, etc.						
PLANT and EQUIPMENT						
Material Hoists						
Personnel Hoists						
Cranes						
Concrete Pump						
Masonry Equipment						
Concrete Equipment						
Compressors						
Water Removal						
Travel Expense						
Room and Board						
Wage Increases						
Total This Sheet C						
Forward From Sheet B						
Forward From Sheet A						
IP-5,000 - 7-70 TOTAL						

FIGURE 4.6.2 *(cont.)*

Units of quantity may not be compatible with your own pricing system, and pricing problems could develop. The purchased data may show for a slab only two items—cubic yards of concrete and square feet to be finished. Your system may show four items—cubic yards, the lineal footage of screeds, square footage of trowel or special finish, and area to receive an application of curing compound. You could end up underpriced if you happen to omit the two missing items.

Conversely, if you generally price only two items—namely, cubic yards of concrete and its finishing—and the estimate shows four items, you may well be overpriced.

Some people insist that construction costs would be lower if all contractors were given quantity surveys along with the plans and specifications. I disagree for the reasons outlined above.

4.8 SUBCONTRACTORS AND SUPPLIERS

Bids from subcontractors and material suppliers are an essential part of your cost estimate. Many general contractors mail out inquiries with reply cards to learn which subcontractors and material suppliers are planning to submit price quotations. On bid day it is too late to be looking for subcontract prices. See paragraph 4.5.4.

Subcontract prices are received on many items of work. It is best to have a form on which sub-bids received by telephone can be recorded. This form, Fig. 4.8.1, has spaces for all the pertinent information: the section covered by the bid, the date, and by whom the bid is received is recorded. (This is important if an error is made and either clarification has to be made or blame assessed.) Next come the name of the firm and the person calling in the bid as well as the telephone number. (This information is important if you have to call back for the clarification of the bid.) The type of work is next and should tie in to the section numbers. Addenda included should be noted to make sure they have a total bid. Whether or not sales tax is applicable, and if it is, has it been included. The price and whether the item is FOB or furnished and installed is also included. In actual use you can fill in, prior to bid day, the name of the job, and addenda numbers that have been added. This saves time and acts as a check on how many addenda were issued and whether or not tax is to be included. This form with the basic items filled in can then be distributed to those who will be receiving sub-bids.

The remarks, of course, would include any exclusions or exceptions that would affect the bid. Alternates could also be listed here either in advance (as you would addenda) or as they are applicable to the particular trade whose price you are taking.

It is important to ascertain before bid day that all the trade work you do not do with your own forces is covered. It sometimes is too late on bid day to find out the item is not covered. If it is a simple item, your staff can quickly come up with a fairly accurate price for the item. However, if it is a complex one or a trade you are not very familiar with, the guess you make could be way out of line and possibly make your total bid too low or too high; neither is a very good position to be in.

Prices on large subcontract items are often the last ones to come. When a price is missing, your chief estimator must get a quick takeoff or make an educated guess. This accomplishes many things. First, you are able to put together a fairly realistic total price on your recap sheet. This allows you to review the square foot cost, and check it against your past experience to see if your bid is in line and realistic. This also allows you time to give thought to whether you want to add a contingency (if

BID PRICE SHEET   Section (s)_____

                         Job_____

                                        Date_____ Rec'd by_____

Firm_____

Address_____

By_____   Telephone_____

Type of Work_____

_____   Price$_____

Addenda_____   Tax included _____

                                        Delivered only_____

                                        Installed   _____

Remarks:

FIGURE 4.8.1 Subcontract Bid Price Form

one is needed) and to evaluate the amount of profit to put on the job. Next, you know which item is simply a guess, and you can plug in the right price at the last minute via an adjustment sheet (see paragraph 4.9.3) when the subcontractor provides it.

It is not unusual that some subcontract items contain variables. One is Miscellaneous Iron, which sometimes covers many unusual units. Bids will come in in which different subcontractors include or exclude some items as well as their erection but not others.

For these items a "spread sheet," which can be any columnar pad, should be used. See Fig. 4.8.2. Each item in the specification section can be listed in the item column and each sub-bidder at the top of the column across. Under the name of the bidder check off the items included in his bid on the left side of the column and the items erected on the right.

At this point you clearly can see which items are not included or not erected by each bidder. If need be, a sum of money can be added to each bid for the items not included or erected which we have shown. In this way a real and complete comparison of these sub-bids can be made, the low bidder determined, and the proper amount inserted in the bid.

This form can be used for any complex section or when many sections are bid as a unit, as sometimes happens with mechanical and electrical sections.

Some firms use a tab system, for example, in the form of a three-ring notebook. Each section of the specification receives a number which is carefully cross-referenced. A sheet is set aside in the book for each tab number. As sub-bids are received, they are entered on the corresponding tab page. The difficulty with this system is that at the tail end of the bid process, when prices are coming in fast and furiously, the system sometimes breaks down.

There are many computer programs that are presently available for estimating and bid listing. Using a spread sheet program, you can devise your own tally sheet for the subcontractors. Once the system is established and programmed, you can put new bid numbers into the item, and the totals will be automatically adjusted to show the change. This is very fast and accurate.

4.9 HOW TO ORGANIZE

The factors going into the bid are easier to identify if a different color of paper is used for each group of items. For example, sheets showing work done by your own forces could be white, overhead pink, subcontracts yellow, and recap green.

Subcontract sheets are in two parts. The first part, Fig. 4.9.A, is used with a job having only a few breakouts or alternates. The second, Fig. 4.9.B, is used for jobs with many breakouts or alternates and is attached to the first sheet, thereby adding the needed columns. Always list allowances on the first sheet, because this will lessen the chance of overlooking them. When using several sheets, carry over totals to the last sheet as shown on the sample.

SPREAD SHEET MODEL

SPREAD SHEET: MISCELLANIOUS IRON SECTION 9

	ARCHITECTURAL IRON		ORNAMENTAL SHOP		IRON WORKS	
SEC 9 ITEM	FURN	INSTAL	FURN	INSTAL	FURN	INSTAL
.3 LINTELS	*	$4,000	*	$4,000	*	$4,000
.4 WHEEL GUARDS	*	*	*	$300	*	$300
.5 STAIRS	*	*	*	*	*	*
.6 WROUGHT IRON RAILS	*	*	*	*	*	*
.7 PIPE RAILS	*	*	*	*	*	*
.8 SHELF ANGLES	*	$3,500	$5,000	$3,500	*	$3,500
.11 DOCK BUMPERS	*	*	*	*	*	$500
.14 WIRE PARTITIONS	*	*	*	*	*	*
.15 SPARK ARRESTOR	*	*	*	$200	*	*
.16 CHANNEL IRON FRAMES	*	$2,800	*	$2,800	*	*
ADD ITEMS NOT INCLUDED		$10,300		$17,800		$8,300
BID PRICE		$98,000		$105,000		$112,000
TOTAL MISC IRON BID		$108,300		$122,800		$120,300
LOW BIDDER IS ARCHITECTURAL IRON						
	FURN = FURNISHED					
	INSTAL = INSTALLED					

FIGURE 4.8.2 Spread Sheet Format

4.9.1 Bid Day

Bid day is often hectic; this is when the "moment of truth" arrives. Having a calm and cool chief estimator is essential. Serious errors can occur at the last minute. Figures may be left out, decimal points dropped or moved over, and columns not properly totalled. Items of work to be performed by your own forces should have been

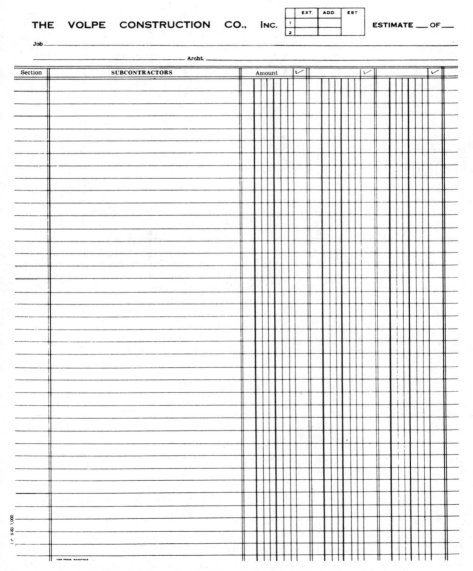

FIGURE 4.9 Subcontract Bid Listing Sheet (continued on next page)

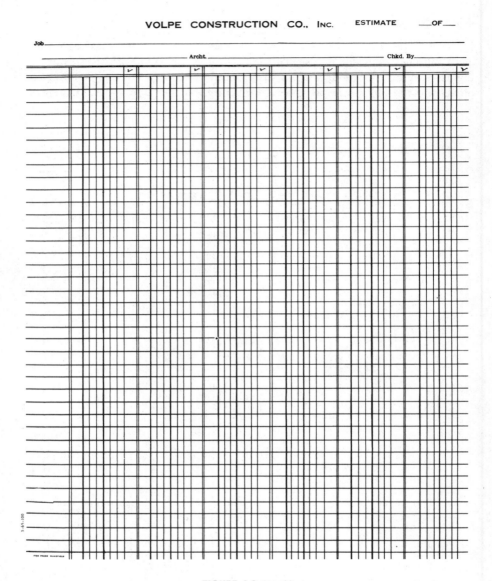

VOLPE CONSTRUCTION CO., Inc.   ESTIMATE   __OF__

Job _____

_____ Archt. _____   Chkd. By _____

FIGURE 4.9 (cont.)

completely priced and recapped. These should be finished the day before bids close to avoid any more confusion than necessary. The profit factor that you intend to add should be clearly established in your mind, whether it be a percentage or a fixed sum. The overhead sheet should also be completed and totaled ahead of time. This leaves bid day free for receiving and evaluating subcontractor bids and arriving at a final price.

THE VOLPE CONSTRUCTION CO., Inc.

	EXT.	ADD	EST
1			
2			

ESTIMATE

Job _____ Bids Close _____

_____ Archt. _____ Cu. Ft. _____ Cost per ft. $ _____

SUMMARY	Material	Labor	
SHEET NO. 1			
SHEET NO. 2			
SHEET NO. 3			
SHEET NO. 4			
SHEET NO. 5			
SHEET NO. 6			
SHEET NO. 7			
SHEET NO. 8			
SHEET NO. 9			
SHEET NO. 10			
SHEET NO. 11			
SHEET NO. 12			
SHEET NO. 13			
SHEET NO. 14			
SHEET NO. 15			
SHEET NO. 16			
SHEET NO. 17			
TOTAL OVERHEAD			
SET MILLWORK LABOR			
TOTAL MATERIAL			
SALES TAX			
TOTAL LABOR			
INSURANCE & TAXES %			
SUBCONTRACTORS			
BOND			
ADJUSTMENTS			
PROFIT .			
TOTAL			
BID			

FIGURE 4.9.2 Bid Summary Sheet Form

4.9.2 Summary Sheet

Estimate sheets (white sheets) and other items making up the costs are listed on the project summary sheet, Fig. 4.9.2. Notice that items for overhead, sales tax, insurance and taxes, all of which belong in the total for work done by your own forces, are shown. Subcontractors, bond, adjustments, and profit are added to reach the total bid price. Some contractors carry the insurance with each classification of work, which may make the insurance add-on more accurate.

4.9.3 Adjustment Sheet

The space for adjustments, on the summary sheet, includes items that are corrections and or last-minute changes. Rather than alter carefully prepared sheets, any late subcontractor prices and revisions can be listed on a separate "adjustment" sheet as shown in Fig. 4.9.3.

PROJECT: ADJUSTMENT SHEET

Item	Amount Carried	Add	Deduct
Painting	$80,000	$10,000	
Plumbing	$1,250,000		$78,000
Fire Protection	$450,000	$35,000	
Heating and Ventilating	$1,890,000		$350,000
Drywall	$450,000		$28,000
Totals		$45,000	$456,000
Deduct			$45,000
Adjustment *Deduct*			$411,000

FIGURE 4.9.3 Adjustment Sheet Form

4.9.4 Final Checking

Serious errors can be prevented by calculating the unit cost per square foot or per cubic foot for the total structure. Space for this is shown at the top of the recap sheet. From past experience you will know if your price is reasonable for the type of project.

Minor errors can still slip by. Both the person doing the original takeoff and the person checking extensions and totals should initial each sheet. This keeps everyone on his toes and helps management identify the best and most accurate performers.

4.10 BID RECORD

Computers are also very useful for saving prices and costs from previous jobs, thereby building a good data base for future estimates. A standard form for this information is shown as Fig. 4.10 and is called a bid record. From these types of records you can get information as to what costs are running per square foot for the various projects you estimate. This information is useful when you are checking your estimate to make sure it is in line and also when you are preparing preliminary estimates. These figures can readily be computerized to make the records easily available and to make quicker comparisons.

TYPE	VOLPE CONSTRUCTION CO., INC. **BID RECORD**	**JOB NO.** _____

DATE BID _____

NAME: _____

LOCATION: _____

OWNER: _____

ARCHITECT: _____

CUBE: _____

FLOOR AREA: _____

NO. OF STORIES: _____

STORY HEIGHTS: _____

BID RESULTS   **NO. OF BIDDERS:**

1. _____ $ _____

2. _____ $ _____

3. _____ $ _____

JAVCCO _____ $ _____

UNIT COSTS

	/CF	/SF
LOW BIDDER	_____	_____
JAVCCO	_____	_____

TYPE OF CONSTRUCTION

SUB FOUNDATION: _____ --

FOUNDATION: _____

FRAME: _____

EXTERIOR WALLS: _____

INTERIOR PARTITIONS: _____

FLOORS: _____

ROOF: _____

CEILINGS: _____

WINDOWS: _____

INTERIOR TRIM: _____

SPECIAL EQUIPMENT: _____

OTHER FEATURES: _____

_____

_____

SEE REVERSE SIDE FOR SKETCH OF SITE, SITE AREA, BLDG. AREA AND OTHER SITE INFORMATION

A/500/4/65

FIGURE 4.10 Bid Record Form (continued on next page)

<u>SITE SKETCH</u>

SITE AREA:	GROUND AREA OF BLDG.

SITE INFORMATION:

EXISTING CONDITIONS:

BORINGS:

UTILITIES:

PAVING:

LANDSCAPING:

OTHER:

FIGURE 4.10 (cont.)

4.11 DATA BASES

We have computerized our cost records for a budget process that allows us to prepare a preliminary budget fairly fast. The program breaks down as per the C. S. I. format of sixteen sections. Each item in the data base has a unit cost established for a specified unit of work. As an example, division 2 sitework is broken down into items such as:

clearing	on a per acre basis,
excavation	on a cubic yard basis
sanitary sewer	on a linear foot basis
storm sewer	on a linear foot basis
fencing	on a linear foot basis
paving	on a square yard basis
landscaping	on a lump sum basis

Each of these items has a per unit cost established in the data base. This applies to other items in the project such as: the skin which is based on a square foot basis and the floor system on a square foot basis. These types of quantities can be estimated very rapidly. The quantities are then inserted into the program which automatically makes the extension and adds it to the estimate. The cost is broken down first by unit estimated then recapped by section. The program shows the total cost for the item as well as the square foot and percentage cost which allows for further checks on your accuracy. This system accommodates different types of construction as well as new or alteration work. The data base can be programmed to include anything you want it to. A copy of page one of a print out is shown. We have deleted the unit, and square foot costs as well as the percentage numbers. The form is included to show what can be done with your computer using this type of program.

4.12 JOINT VENTURES

When participating in a joint venture the parties generally agree that each participant will take off and price the quantities for the work category that the venture will do with their own forces. A comparison is then made to see if there is agreement on the quantities and dollars. If a difference does occur then they must study each estimate in detail to find out where the differences occur and reconcile these differences. Adjustments are then made so that all parties are satisfied. See Chapter 13 for more details on joint ventures.

4.13 CONCLUSION

Successful growth of your firm will depend on bidding projects you are sufficiently familiar with to be competitive and experienced enough in to be profitable. Steady and systematic bidding of projects will keep your estimators up to date and will help maintain a backlog of work on your books. Remember, a steady backlog gives you the luxury of not having to take work at too low a bid, thereby preventing you from having too many losses, which could be disastrous.

| BUDGET #| 1 | BUDGET DATE | 6/1/88 | PAGE 1 |

PROJECT#	PROJECT NAME	LOCATION	JOB DATE	OWNERS BUDGET	ESTIMATED COST	DEVIATION
1	SPV OFFICE BLDG	MALDEN, MASS	9/1/88	$6,000,000	$5,350,339.00	$649,661.00
# STORIES	TYPE OF FRAME	TYPE OF SKIN & WINDOWS	BLDG. AREA			
7	CONC	BRICK & WINDOWS	67,500			
		SITE AREA				
		182,500				

DIV 1. GENERAL REQUIREMENTS	$600,000	DIV 9. FINISHES		$319,480
DIV 2. SITE WORK	$145,400	DIV 10. SPECIALTIES		$23,200
DIV 3. CONCRETE	$1,309,469	DIV 11. EQUIPMENT		$15,000
DIV 4. MASONRY	$221,000	DIV 12. FURNISHINGS		$34,850
DIV 5. METALS	$259,500	DIV 13. SPECIAL CONSTRUCTION		$0
DIV 6. WOOD & PLASTIC	$125,000	DIV 14. CONVEYING SYSTEMS		$270,000
DIV 7. THERMAL & MOISTURE PROTECTION	$59,900	DIV 15. MECHANICAL		$1,075,500
DIV 8. DOORS & WINDOWS	$284,540	DIV 16. ELECTRICAL		$607,500

DIV	CODE	UNIT DESCRIPTION	UNIT	QUANTITY	UNIT COST	TOTAL COST	S. F. COST	% COST
2	2100	CLEARING	ACRES	4				
2	2200	EXCAVATION	C. Y.	6000				
2	2570	SANITARY SEWER	L. F.	200				
2	2580	STORM SEWER	L. F.	200				
2	2710	FENCING	L. F.	1000				
2	2610	PAVING	S. Y.	1200				
2	2800	LANDSCAPING	L. S.	1				

FIGURE 4.11 Preliminary Budget

Types of Contracts

Lump Sum

Lump Sum Short Form

Government

Cost Plus a Fee

Guaranteed Maximum Plus A Fee

Construction Management

Construction Management with a Guaranteed Maximum Price

Project Management or Administration

Unit Price

Nonstandard

Design–Build

There are many different types of contracts used in the construction industry. Each type fits the particular situation that best serves the owner's purpose, from either a financial or time factor point of view. The following paragraphs explain the different types of contracts and the advantages and disadvantages of each.

Standard forms are available for the various types of contracts. These should be used because they have gone through many years of experience as well as the courts, which makes the articles better understood. Their use and end results are therefore predictable. General conditions are also available and should also be used. Modifying either the contracts or general conditions should be kept to a minimum in order to avoid potential changes to the aforementioned benefits. A new clause may also alter the meaning of some of the other clauses in the contract and or cause conflicts. It takes a lot of study and cross checking to make sure this does not occur. Special and supplemental conditions should not add conflict to the standards either.

Another factor to be considered is the cost of legal fees. When a contract is modified to a great extent, both parties have to expend considerable sums of money to make the contract agreeable to both parties. It is difficult to write the perfect contract but people still try, with dubious benefit and at considerable expense.

5.1 LUMP SUM CONTRACTS

Lump sum contracts are generally used for bid work whether it be private or public. The owner, usually through his architect, has plans and specifications prepared. These plans should be 100% complete, and in essence are working drawings ready for the construction of the project. This allows bids to be taken from the general contractor based on the finished plans and specifications. All bidders know what is expected and are all bidding on the same basis. The contract is then awarded to the lowest responsible bidder. The contract most generally used on private work is A.I.A. Form A101 which covers the scope of work, completion time, progress payments, and acceptance of the work. This basic form has been used for years, being periodically updated from time to time.

General Conditions A.I.A. Form A201 is used in conjunction with the A.I.A. Form A101 as well as many other A.I.A. contract forms. This too has been in existence for many years and has been updated periodically and is understood by both parties to the contract. Sometimes supplemental or special conditions are added to A.I.A. General Conditions A201 to take care of some special factors that may be involved in the project.

In preparing bids for these types of contracts you should estimate the construction costs for the work you do with your own forces. The general conditions costs will vary with each project, and it is important to review the general, special and/or supplemental conditions to make sure all the items required are priced. Some people use a percentage of the job cost for general and special conditions, but these costs should actually be estimated because a percentage figure is not accurate enough on larger jobs.

In lump sum contracts you must be accurate with your estimate. You are paid a fixed sum of money to do this project. If your estimate is too low, you will obviously lose money. If your estimate is too high, you are not likely to be low bidder and therefore will not get the job. It costs money to bid on work and it does not make much sense to invest your time and effort unless you are serious about trying to get the job.

5.2 LUMP SUM FOR SMALL CONTRACTS

For smaller contracts the A.I.A. Form A107 is generally used. It is a self-contained-contract with general conditions included. It is easily understood and again tested by time and the courts. Only the basic facts and figures have to be entered to complete this contract form. It is very convenient for both parties to use this form because of its simplicity.

5.3 GOVERNMENT CONTRACTS

Federal and state agencies and their various subdivisions use these contracts which vary with the agencies. Some have standard general conditions. States have different forms, generally none being the same. Counties and municipalities also have their own form of contract with general conditions that are designed for their use. States, counties, and municipalities all have varying laws that get incorporated into the contracts. As a result, each one has to be evaluated on a contract by contract basis. Your attorney should be consulted on these types of contracts. Some of the conditions can seem simple but cost a lot of money to implement. In public work the contract form used varies with the governmental agency, state, or subdivision. Some may use the same form. The military uses one type of form; the General Service Administration and the Veterans Administration use others. Generally speaking, many of the clauses are similar. In public work many references are made to various government specifications that are not always incorporated in the specification. These can be obtained from the agency involved and you should read them to familiarize yourself with their contents to understand what is called for. Most follow standard procedures, but some are more exacting: this must be taken into account during the bidding and construction phase.

5.4 COST PLUS A FEE CONTRACTS

This form of contract is very useful when the exact extent of the project is unknown. Sometimes a project must get started before plans are finished and a firm price cannot be ascertained. A building damaged by fire or some other disaster with unknown structural damage is another instance where this type of contract is useful. A.I.A. Form 111 may be used for this type of contract.

Contained in this form of contract are spaces where you list the type of fee you are to be paid, whether it is a percentage of the final cost of the project or whether it is a lump sum amount. Also there is a list of the costs that are reimbursable such as materials, labor, rented equipment, and insurance. There is also a list of the nonreimbursable items that are covered by your fee. These lists should be reviewed to make sure this is your understanding. If you expect to be reimbursed for some people not stationed in the field, this contract does not allow it, unless you will have reached an understanding that will be added to the contract to include it as a reimbursable cost. Insurance, permits and taxes should be reviewed to make sure you are aware of what is reimbursable and what items are covered by your fee.

5.5 GUARANTEED MAXIMUM PLUS A FEE CONTRACTS

This is an extension of the cost plus a fee contract except that it puts a cap on the cost to the owner. The contract lists all the items of cost to be reimbursed including the general condition items. The fee is spelled out either as a percentage or a lump sum.

This type of contract is generally used in a negotiated situation where a price and a fee are established after reviews with the owner and architect of the work to be done. A.I.A. Form A111 may be used for this type of contract adding the guaranteed maximum sum feature to the contract.

The costs are reimbursable as in the cost plus a fee contract unless the project costs more than the guaranteed maximum price (GMP) then the contractor absorbs the overrun. If the costs run less than the GMP then the owner saves this amount of money. In most of these contracts there is an incentive added in which you participate in the savings. The savings split is anywhere from 50% to the owner and 50% to the contractor to 75% to the owner and 25% to the contractor. These percentages are negotiated as the contract is being finalized. This type of contract must keep track of all changes and change orders issued to increase or decrease the GMP as the change order provides. If this is not done, the contract can overrun because of these changes and you will be hurt. It also could affect the savings that you are to participate in. Accurate records must be kept because the owner generally has the right to audit the job records to ensure that all items charged to the project were proper. This option is generally open for two or three years.

5.6 CONSTRUCTION MANAGEMENT CONTRACTS

Construction management has several variations, which is why some people get confused as to its use and meaning.

Generally speaking, the owner selects the architect and construction manager almost simultaneously. He may hire one first and have that party help select the third, or he may select both without help from the other two parties. Whomever he selects should be compatible with each other to ensure cooperation throughout the project.

The A.G.C.A. has a good booklet out entitled "Construction Management Guidelines." This booklet is handy when determining the qualifications needed and the knowledge possessed by each potential construction manager.

A.G.C.A. booklet entitled "A Guide to Owner's Responsibility for Construction Projects" is also very helpful to owners in understanding what is expected of them during the various phases of the project. Some owners are not sure what construction management is and this booklet is very useful to them.

First the "team," owner, architect and construction manager, meet to review the owner's needs. A budget is prepared either from his program needs or schematic drawings. If the budget is within the owner's appropriation, the design can begin. If not, the program or schematic drawings have to be changed to meet the owner's appropriation. Even during the design development stage, the team can continue to value engineer the project to make sure the owner gets the most for his money and keep it within an affordable budget. This monitoring continues all during the plan development stage to ensure that the costs stay within the budget. The greatest service performed by a construction manager, in my estimation, is during the design phase working with the architect/engineer and the owner to ensure the project will stay within the owner's budget and that the project is buildable within the allotted time frame.

During the construction phase the construction manager prepares bid packages for each of the trades required. These packages are assembled to prevent overlap or voids in the trade contracts. Trade jurisdictions are also evaluated in arriving at the packages. Allocating general condition items is another function in assembling the packages.

The construction manager then takes bids for the various trade items. The trade contractors may be prequalified if so desired by the owner or the construction manager. Bids are reviewed by the construction manager after which he makes a recommendation to the owner for award.

The construction manager supervises the project including quality control, checking the time schedule, reviewing change orders, and keeping the owner and architect/engineer informed if a potential overrun may exist. He runs the job meetings, keeping and distributing the records of same to the parties involved. All correspondence to or from the trade contractors should always be through the construction manager to avoid confusion on the project.

The construction manager and the architect determine when substantial completion occurs. He works with the architect to make up the punch list of small items still to be completed.

At the close-out he ensures that all the warranties and guarantees as well as "as-built drawings" and manuals are assembled and turned over to the owner.

Construction management can be done on a no-risk basis where all of the construction manager's costs are reimbursed and there is no guaranteed price, or on a guaranteed maximum price basis.

When acting as a pure construction manager, the likelihood of an adversarial relation with the owner is almost nil. When a guaranteed maximum cost is added, this of course can potentially create some adversarial relations.

In some contracts the construction manager may do some of the work with his own forces whereas in others he does only the general condition items. A clear understanding should exist as to the basis of what work is to be done by the construction manager's own forces. It can be done on a lump sum basis or cost plus a fee. You in essence become a subcontractor to yourself so you have to be careful to keep the people working on the subcontract separate from the construction manager contract. This ensures that the owner receives fair treatment from this arrangement.

The Associated General Contractors of America has construction management contract forms for either the pure or the guaranteed maximum feature.

5.7 PROJECT MANAGEMENT FORM OF CONTRACTS

Project management contracts go one step farther than construction management. Under this form of contract you act for the owner from start to finish of the project. You help select the site, establish his budget, select the architect and the construction manager or general contractor, and supervise the project through completion for the owner. The actual turnover of the project includes training plan facility people in the operation of the building equipment. You make sure the punch list is completed as agreed and gather the warranties, guarantees, manuals, and as-built drawings to be delivered to the owner. This form of contract can also be called project administrator. Although the title "project manager" is used quite often we prefer to use the term project administrator. This terminology avoids the confusion that may occur when talking about the person who acts as a project manager for a specific part of the program.

We have noticed recently that some are calling this form of contract "program manager." Included in the appendix is a form of contract we use for project administrator. This form should be reviewed by your attorney and insurance carrier. It will also have to be reviewed for each project involved as responsibilities may vary with different types of projects as well as different owners. Some may want you to assume more responsibility while others expect you to assume less responsibility.

5.8 UNIT PRICE CONTRACTS

This type of contract is used where quantities are easily measured and can be pretty well defined beforehand.

A good example is when an office building is complete and tenant work has to be done. The amount of partitioning cannot be determined until the spaces are leased. In this case unit prices are established for each lineal foot of the different types of partitions. Unit prices for various types of doors, electric outlets and switches, different types of ceilings on a square foot basis, etc., are established. The unit price includes your cost plus an amount for overhead and profit. See Fig. 5.8.1 pages 1 and 2 for an illustration. On some larger buildings you may have up to 200 or more unit prices. These should be set up on a computer to make it easier to keep track of and to prepare payment requisitions. When the tenant layout plans are ready the quantities of items required are taken off and multiplied by the unit price established to come up with a cost for the area involved.

Some contracts for public works such as highways are priced on a unit price basis. The proposals give all the items involved and are listed as units with a quantity. You then price each item with a unit price that includes your overhead and profit. The item quantity is then multiplied by your unit price to arrive at a total for each item. These are then added up to get to the total for the bid. The contract is then awarded to the low bidder based on this total.

When the work is being done, the quantities are measured and the payment based on the actual quantities put in place. Most contracts include a clause that if the quantities run over or under by a certain amount, usually 15%, the unit price is adjusted. The reason is that if you have a large reduction of an item that uses equipment which has a special set up cost in your bid, and you have averaged this cost into the total quantity in the estimate, and if a lesser quantity was installed, your unit cost would be too low, and you would lose money. On the other hand, if the quantity increased substantially, your average cost would drop. You will note some

TENANT WORK UNIT PRICES

M O B TENANT WORK				DATE
SUITE # 215				10/14/89
ITEM	QUANT	UNIT	UNIT COST	TOTAL
DRYWALL	INST			
CORRIDOR PARTITION - 1 LAYER BOARD		L F	$26.50	
DEMISING WALLS 3 5/8" STUDS W / BATTS		L F	$61.00	
INSIDE FACE OF EXTERIOR WALL		L F	$8.25	
TENANT PARTITIONS 6" STUD - 3" INSULATION		L F	$52.10	
PARTITIONS WITH INSULATION		L F	$46.30	
CHASE PARTITIONS		L F	$30.25	
PARTIAL HEIGHT PARTITION 42" HIGH				
(WITH 12" DEEP COUNTER TOP CAP)		L F	$71.00	
COLUMNS (INTERIOR)		EA	$240.50	
COLUMNS (EXTERIOR)		EA	$145.75	
MULLION CAPS (BETWEEN MULL & COL /WALL)		EA	$48.40	
DROPPED SOFFIT		SF	$10.00	
BLOCKING		L F	$2.75	
CEILINGS		S F	$4.48	
DOORS (INCLUDING FRAME, FINISH & HARDWARE)			$0.00	
ENTRY DOOR W/SIDELIGHT CUT INTO CORRIDOR		EA	$1,810.00	
INTERIOR DOORS		EA	$710.40	
MILLWORK				
PLAIN WINDOW SILL 3/4" X 7"		L F	$19.30	
BASE CABINETS W/ COUNTER TOP		L F	$145.00	
WALL CABINETS		L F	$110.00	
SHELVING, BRACKETRY AND STANDARDS		L F	$27.10	
COAT RACK - RED OAK PANEL WITH HOOKS		EA	$126.00	
DRAWER UNITS		EA	$341.00	
PLASTIC LAMINATE BENCH AND DOOR HOOKS		EA	$152.00	
COUNTER GRILLE		LS	$110.00	
PLASTIC LAMINATE COUNTER W/BKSPL & LEGS		LF	$92.50	
WALL COVERING				
VINYL TYPE 4		S F	$2.40	
VINYL TYPE 5		S F	$2.50	
FABRIC TYPE 1		S F	$2.70	
ACOUSTICAL CEILINGS		S F	$3.00	
FLOORING				
CERAMIC TILE FLOOR		S F	$9.50	
CERAMIC TILE BASE		L F	$12.00	
MARBLE THRESHOLDS		EA	$45.00	
SHEET VINYL		S F	$3.40	
RUBBER WALL BASE		L F	$2.10	
CARPET		SY	$24.00	
VERTICAL BLINDS		EA	$16.00	
TOILET ACCESSORIES		EA	$440.00	
TOILET ACCESSORIES - HANDICAPPED		EA	$495.00	
H V A C				
SUPPLY DUCT				
26" X 14" LINED		L F	$83.50	
26" X 12" LINED		L F	$81.10	

FIGURE 5.8.1 (page 1)

TENANT WORK UNIT PRICES

HVAC CONT.			
6" X 6" UNLINED	L F	$19.12	
8" X 8" UNLINED	L F	$24.25	
10" X 4" UNLINED	LF	$20.06	
10" X 8" UNLINED	L F	$25.62	
10" X 10" UNLINED	L F	$27.40	
12" X 8" UNLINED	LF	$46.20	
12" X 10" UNLINED	LF	$45.10	
14" X 10" UNLINED	L F	$48.00	
18" X 10" UNLINED	LF	$54.10	
24" X 10" UNLINED	L F	$60.50	
24"X 12" UNLINED	L F	$51.25	
RETURN DUCT			
22" X 18" LINED	LF	$80.00	
22" X 16" LINED	LF	$78.40	
TRANSFER DUCT 20" X 10"	EA	$500.00	
" " 12" X 10"	EA	$350.00	
" " 10" X 10"	EA	$350.00	
TOILET EXHAUST DUCT & GRILLE	L F	$14.50	
SUPPLY DIFFUSER W/5' FLEX	EA	$180.00	
RETURN AIR GRILLE	EA	$88.40	
FAN COIL CONTROLS, WIRING	EA	$850.00	
DUCT HEATER	EA	$1,475.00	
COOLING THERMOSTAT	EA	$300.00	
PLUMBING, COMPLETELY PIPED			
TOILET	EA	$2,510.00	
LAVATORY	EA	$2,250.00	
COUNTER SINK	EA	$2,250.00	
LAVATORY, HANDICAPPED	EA	$2,700.00	
TOILET, HANDICAPPED	EA	$3,010.00	
ELECTRIC WATER HEATER (20 GAL)	EA	$1,200.50	
FIRE PROTECTION - HEADS	EA	$310.50	
ELECTRICAL			
LIGHT FIXTURES 2' X 2'	EA	$175.25	
" " 2' X 4'	EA	$185.40	
HI-HAT LIGHT FIXTURES	EA	$200.00	
UNDER CABINET LIGHT	EA	$180.40	
EXIT LIGHT	EA	$350.50	
EMERGENCY LIGHT	EA	$325.00	
SINGLE POLE SWITCH	EA	$88.50	
THREE-WAY SWITCH	EA	$110.40	
TELEPHONE OUTLETS (EMPTY)	EA	$40.10	
COMPUTER OUTLETS (EMPTY)	EA	$40.10	
BASEBOARD RADIATION	L F	$45.75	
ELECTRICAL WALL HEATER	EA	$240.50	
POWER			
RECEPTACLES	EA	$92.10	
RECEPTACLES - DEDICATED	EA	$185.20	
GFI RECEPTICLES	EA	$140.50	
PANEL BOARD & SERVICE	L S	$8,000.00	
TOTAL INSTALLED			

FIGURE 5.8.1 *Tenant Work Unit Price Form (page 2)*

items are lump sum units. Manholes, for instance, are grouped in the various sizes required. This avoids breaking this type of item into unit prices for each part of the manhole thereby reducing the number of unit prices to be bid. See Fig. 5.8.2 for three sample sheets from an actual bid form. Included in the bid form is a description of the items being bid which gives you the information needed in order for you to know what is included in each unit bid item. As an example:

ITEM 120.1   UNCLASSIFIED EXCAVATION   CUBIC YARD
The work under this Item shall conform to the relevant provisions of Section 120, and the following: Unclassified Excavation shall consist of the excavation and satisfactory disposal of all materials encountered within the limits of contracts, except those classified and paid for under other items.

ITEM NO.	QUANTITY	ITEM WITH UNIT BID PRICE WRITTEN IN WORDS	UNIT PRICE		AMOUNT	
			DOLLARS	CENTS	DOLLARS	CENTS
101	0.6	CLEARING AND GRUBBING AT_____ PER ACRE				
103	5	TREE REMOVED — DIAM UNDER 24 IN. AT_____ EACH				
104	2	TREE REMOVED - DIAM 24 IN. AMD OVER AT_____ EACH				
113.1	125	DEMOLITION OF SUBSTRUCTURE AT_____ PER CUBIC YARD				
114.1	1	DEMOLITION OF SUPERSTRUCTURE (BRIDGE NO. P-10-63) AT_____ LUMP SUM				
120.1	6,400	UNCLASSIFIED EXCAVATION AT_____ PER CUBIC YARD				
140.	320	BRIDGE EXCAVATION AT_____ PER CUBIC YARD				
144	5	CLASS B ROCK EXCAVATION AT_____ PER CUBIC YARD				
		CARRIED FORWARD				

FIGURE 5.8.2 Unit Price Bid Form (continued on next page)

ITEM NO.	QUANTITY	ITEM WITH UNIT BID PRICE WRITTEN IN WORDS	UNIT PRICE DOLLARS CENTS	AMOUNT DOLLARS CENTS
150	4,500	ORDINARY BORROW AT_____ PER CUBIC YARD		
151	2,412	GRAVEL BORROW AT_____ PER CUBIC YARD		
156.1	190	CRUSHED STONE FOR BRIDGE FOUNDATIONS AT_____ PER TON		
170	6,100	FINE GRADING AND COMPACTING - SUBGRADE AREAS AT_____ PER SQUARE YARD		
270.481	170	TEMPORARY 48 INCH PIPE AT_____ PER LINEAR FOOT		
402	300	DENSE GRADED CRUSHED STONE FOR SUB-BASE AT_____ PER CUBIC YARD		
420.	425	CLASS I BITIMINOUS CONCRETE BASE COURSE, TYPE I - 1 AT_____ PER TON		
443	7	WATER FOR ROADWAY DUST CONTROL AT_____ PER 1000 GALS		
		CARRIED FORWARD		

FIGURE 5.8.2 (cont.)

Included under this item shall be the removal and disposal of Bituminous Pavements, Paved Water Way, Bituminous concrete Curbs, and any other material not included under any other item. Also included is the removal of all stone, earth, and pavement materials used in the temporary detour road. The materials shall become the property of the contractor, and all disposal shall be off site.

5.9 NONSTANDARD CONTRACTS

A nonstandard form of contract is a type that some of the larger corporations use. The contract is usually very wordy and contains some onerous clauses. These contracts have evolved from the standard forms, and one new clause after another has been added. The final document has no resemblance to any standard form. These types of

ITEM NO.	QUANTITY	ITEM WITH UNIT BID PRICE WRITTEN IN WORDS	UNIT PRICE		AMOUNT	
			DOLLARS	CENTS	DOLLARS	CENTS
851	1	SAFETY CONTROLS FOR CONSTRUCTION OPERATIONS AT_____ LUMP SUM				
852	200	SAFETY SIGNING FOR CONSTRUCTION OPERATIONS AT_____ PER SQUARE FOOT				
853.2	100	TEMPORARY PRECAST CONCRETE MEDIAN BARRIERS AT_____ PER LINEAR FOOT				
854.014	2,600	TEMPORARY PAVING MARKINGS - 4 IN (PAINTED) AT_____ PER LINEAR FOOT				
859.1	20,000	REFLECTORIZED DRUM WITH FLASHER (TYPE A) AT_____ PER DRUM DAY				
860.0	1500	4 IN. REFLECTORIZED WHITE LINE (PAINTED) AT_____ PER LINEAR FOOT				
861.04	1500	4 IN. REFLECTORIZED YELLOW LINE (PAINTED) AT_____ PER LINEAR FOOT				
874.3	1	TRAFFIC SIGNS REMOVED AND STACKED AT_____ LUMP SUM				
		CARRIED FORWARD				

FIGURE 5.8.2 (cont.)

contracts must be carefully studied before making your final bid. Insurance clauses must be checked by your agent because special limits are sometimes added that can be either expensive or unobtainable. Your lawyer should also review the contract to ensure you understand what you are agreeing to.

5.10 DESIGN BUILD CONTRACTS

A Design-build contract is one in which you contract to do the design work (architectural and engineering) as well as the construction of the project. Some firms have in-house design capabilities and others hire an architect/engineer to do the design for the project under their direction. This type of contract adds a great deal more responsibility to your firm. If you have in-house people then you know the risks involved. If you hire an outside firm then you must make sure they have adequate errors and omissions insurance to protect them as well as your firm in case a problem develops. Check with your carrier because additional coverages may be needed in your coverages when you do this kind of work. We would also suggest checking with your attorneys.

5.11 CONCLUSION

These are the basic forms of contracts. There can be variations to all that have been enumerated. When you study the specifications, make sure you check the type of contract to be signed whether it be for a private or public owner. I repeat that if it varies from the standard, by either deletions or additions or a completely nonstandard contract, have your insurance carrier and attorney review it so you can allow for any additional monies for special insurances or contingencies in order to protect yourself. You will note that on the cover sheet of all standard contracts there is a suggestion that you consult your attorney and insurance company because of legal and insurance consequences.

The Associated General Contractors of America and the American Institute of Architects have many Publications and Services available for use. Included as Appendix A are five pages of the A.G.C.A. catalogue "Publications and Services," which includes a listing of some of the contract forms available not only from A.G.C.A. but from the A.I.A. as well. Most of the forms by A.G.C.A. were developed by members on committees charged with that responsibility. The members develop these documents from firsthand knowledge of our contracting practices and try to make these forms as equitable as possible for both parties.

CHAPTER 6

Owner–Architect–Engineer Relations

6.1 WE ARE A SERVICE ORGANIZATION

In the construction business we perform a service. Anyone who has a wheelbarrow or hammer and saw can be classified as a general contractor when he contracts to handle the many phases of an entire construction project from a small remodeling job to a house or whatever. However many cities, counties and states have licensing requirements for various classifications of work. Some licenses are for contractors, others are for superintendents and some are for a specific item of work. The requirements of the area where the work is to be performed must be researched to find out what is required. Do not try to do work in an area unless proper licenses are procured. The results could be devastating, the least of which is that payments may not be made and a lawsuit cannot be brought by the firm to try to recover. The firm can also be fined for operating without a license. If you are going to operate in a state other than the one where the firm is registered, there are laws that have to be researched such as those governing registration as a foreign corporation.

What do we really do? We take an idea—our own, an architect's, or an owner's—and translate it into reality, in other words, a new structure of some type. In this way we are performing a service.

6.1.1 Continuous Service

Each client is entitled to our best performance. If opinions on the plans or the project are requested during the bid preparation or the award phase, honest answers must be given. You are also entitled to honest answers if you have questions during the bid or construction phase.

Once construction begins, the owner and architect are entitled to everything in the specifications and no less. After completion, there are obligations during the guarantee period. Servicing an account is a continuous, not a hit-or-miss operation. Good service is a must for your reputation and repeat business.

6.1.2 Quality

Quality is a simple word, but to some people it means minimal and others maximum. The owner and architect expect a good quality of workmanship and materials, even in a warehouse. I do not mean to demean a warehouse but only refer to it as a less complicated structure.

Specifications for materials sometimes produce a result that makes it seem that the contractor has not done quality work. For example, an "economy" grade of windows may be specified in order to meet a budget. These windows may allow a certain amount of air leakage. If the owner is unhappy with the results he feels cheated. Such a complaint may not be easy to resolve, but when all facts are brought out, you should be in the clear. In this type of situation the relative differences in material must be made clear at the outset. When a lesser quality product is used to make a budget, the owner must be made aware of what he is buying by pointing out the differences in the performance of the items. He may not want to make the change and pay the extra money to get the better quality product. If he does not know what he is getting then he will certainly complain about the product if it does not produce the result he expected.

Windows are just one example; the same problem can occur with numerous other materials that go into a building. As a general contractor, it is your obligation to meet specifications. When you do not, there goes that reputation!

Sometimes it saves money to furnish better materials than specified. A case in point is wood framing. A better grade of lumber is less costly to erect because of fewer knots, straighter pieces, and less splintering. Weigh the added cost of better materials against savings in labor. When erected cost is not greater, here is a way to build goodwill at no expense. If you save money so much the better.

6.1.3 Timely Completion

On-time completion of a building is vitally important to the owner. He has plans for moving, specific dates for installing equipment and machinery, and possible increases in personnel. A delay or postponement can be very costly to him.

The last coat of paint and tile on the floor are what the owner sees when he walks into his new building at scheduled completion. If these are not complete, or have been done poorly, the fact that the best materials were used throughout becomes meaningless. Your entire effort for reputation and good relations could go down the drain.

Legitimate delays for time extensions can and do occur. It is not a good situation, but be sure to give notice to the architect and owner as soon as it becomes apparent that there will be a delay in delivery. Keep them informed so plans for moving can be altered at the lowest cost.

Delays can sometimes be overcome by spending additional money for overtime labor on a selective basis or readily available materials at higher cost. On-time or earlier completion can save on your overhead cost and may help you avoid part of a heating-season expense. Your superintendent is freed for another project. Weigh the added costs against savings and make the decision as to which route to go.

6.2 PRELIMINARY BUDGETS

An architect or good client may ask you to prepare a preliminary cost budget from a program with square footages for different types of areas. The budget may be prepared from schematic drawings or partial or completed drawings. Sometimes you are reimbursed, other times you do it as a service. You gain three advantages from this effort. First, you get a good insight into the details of the project. Second, with a good preliminary budget, bids are likely to be in line and the project will go ahead thereby eliminating costly rebids. Third, you create additional goodwill with the client and architect and may have an opportunity to negotiate the contract for the project. A good cost system with the costs determined from either a square foot basis or actual quantity basis makes the preparation of these estimates easier. The costs you have saved from previous jobs or estimates should be placed into a good data base system as mentioned elsewhere in the book. This makes your estimating units easier to retrieve and you get a more accurate budget because of the dependability of these unit costs.

6.3 DEALINGS WITH ARCHITECTS, ENGINEERS, AND OWNERS

Certain architects, engineers, and owners have the reputation of being tough to deal with. Many times this reputation is caused by poor performance on the part of some past contractor.

This does not have to be. If you carry out a project as specified and do not try to take advantage of a situation, this tough attitude seldom occurs.

Personality clashes may occur at the job site. Be alert to this possibility, and settle things before a mountain develops from a molehill.

When discussing an estimate or change, do not hesitate to show your figures. The open book method is the easiest way to do business. The architect and engineer have some knowledge of costs, and a look at your figures helps remove doubt from the minds of both.

Trust is very important in the construction business in both the public and private sector. Authority held by an individual will vary, and you must know the limits. For example, on some government contracts, the resident engineer can authorize a change if it does not exceed a certain dollar value. If he oversteps his authority, you could do work and not get paid. Similarly, there are some appropriations on public work that cannot be exceeded. An agency might authorize more work and there would be no funds available to pay you. These facts must be ascertained before beginning the project or added work.

6.4 MAINTAINING PROPER CHANNELS OF COMMUNICATION

Channels of communication must be kept open or chaos will prevail. Starting at the top of a proper channel is the owner, followed by the architect/engineer, general contractor, subcontractor, and material supplier.

Problems develop when proper channels are bypassed. For example, a reinforcing steel supplier believes that a concrete beam is under-designed and brings this directly to the attention of the structural engineer. He agrees and redesigns by adding more steel and making the beam deeper. This change is unknown to the contractor, and he builds forms to the original size. The steel arrives on the job and will not fit into the formwork. After much ado and many telephone calls, the facts about the change come out. The result is delay and added cost. The form has to be torn apart and reconstructed. Instead of a minor change order, you have a major one. As a second example, a supplier gets approval from the owner to provide a newer model of mechanical equipment. The contractor and mechanical subcontractor have not been informed of the change. Neither the roughed-in space nor the connections fit, and more costly problems must be solved.

These types of problems cost money and the question of responsibility for the cost seems to take forever to resolve.

Strained relations will occur when the contractor goes directly to the owner without informing the architect or engineer. Also, when a subcontractor needs to discuss problems directly with the architect or engineer, it should be done with the knowledge of the general contractor. The line of communications should be like a ladder. No step should be skipped going up or down. When you miss a step, look out! It can be expensive and disastrous.

6.5 CHANGE ORDERS

Contrary to popular belief, change orders do not result in a lot of profit for a contractor. Most contractors would rather have a project proceed without them. A change order can slow the momentum of the job; it can also disrupt the schedule and cause other problems. However, perfect jobs without changes are few and far between. Change orders are more fully reviewed in Chapter 7.

6.6 CONCLUSION

Communication through channels will help you maintain good relationships with architects, engineers, and owners. You will avoid unexpected costs and difficulties and gain the reputation of a true service organization.

We do not know a contractor who has ever built a sizable project without change orders. It is essential to have a good workable system for handling all necessary paper work, with responsibility for each action assigned to a specific individual.

It is very important to have a smoothly running project. This is accomplished by having all the problems brought out up front and resolved as soon and as amicably as possible. When you finish a project your reputation is further enhanced. If you and the other parties wind up in court or arbitration it costs all parties a lot of money that could have been better spent resolving the issues at the outset and getting them all behind you.

CHAPTER 7

Accounting and Cost-Keeping Systems

7.1 THE NEED

Accounting is a very important facet of any business and the construction business is no exception. Figures on the financial condition of a firm should be available at all times for bankers, bonding companies, and your own everyday operations. Do not skimp on setting up a proper system of accounting.

Cost of operating the firm, generally identified as office overhead, must be known because this overhead expense is spread among the projects under construction when calculating the profit. You can't afford to guess what your net will be—you must *know.*

7.2 THE AUDITOR

Smaller firms generally have an employee who does bookkeeping along with other duties. An auditor comes in periodically and checks the books to make sure accounts are in proper order. The status of receivables requires continuous attention. Overdue accounts are flagged and followed up. The auditor will show how to keep current with payables so that you maintain a good credit rating. Payroll deductions for taxes, social security, and unemployment compensation must be computed accurately and payment made to the proper authorities in a timely manner. There are penalties when these are not handled properly.

The auditor's regular checks also helps avoid unauthorized use of company funds or mismanagement of same. It ensures proper posting to accounts, thereby accurately reflecting their status.

In larger firms, a full-time accountant handles the full range of accounting functions. This does not eliminate the need for an outside auditor on an annual or semiannual basis. Your in-house audits must be confirmed by a certified public accountant to satisfy banking and bonding institutions. When the firm is publicly owned, the outside audit is needed for annual statements.

Figures from your in-house accountants should include a full statement on a quarterly basis with detailed figures on each current project. This information coupled with your general expenses shows whether you are making or losing money.

7.3 USE OF COST-KEEPING SYSTEMS

Cost-keeping records are the backbone of knowledge for learning the financial condition of each project. Records show daily or weekly where your costs stand against estimated costs. The system can be as simple or complex as you want. By starting with a simple system, you can add detailed information as the need develops. Whether you do a little or a lot of field work with your own forces, a cost system is the basis for keeping track of current projects and bidding future projects. There are two cost records that have to be kept. The first is for the work you do with your own forces. The second is for the total project.

7.4 KEEPING COSTS ON THE WORK YOU DO WITH YOUR OWN FORCES

A system of cost keeping is a must and you can develop one to fit your needs. The type of work you do will govern the final system you use. The system shown below is based on the Construction Specification Institute 16-division breakdown; some of the items are shown here to give you an idea how they are listed.

01000   GENERAL REQUIREMENTS

    01   050 Salaries
    01   300 C.P.M./Photos
    01   400 Testing/Inspection/Permits
    01   500 Temporary Facility/Offices
    01   501 Temporary Utilities
    01   502 Plant and Equipment
    01   503 Barriers
    01   504 Security
    01   505 Safety
    01   506 Cleanup
    01   507 Winter Conditions
    01   510 Insurance

02000   SITE WORK

    02   110 Demolition
    02   375 Underpinning

02200		EXCAVATION

	02	201 Hand Exc & Backfill Buildings
	02	203 Gravel under Slab on Grade
	02	204 Fine Grading

02400		DEWATERING

Concrete work is complex and cost keeping is easier when each item of work is listed separately.

03100		CONCRETE FORM WORK

	03	101 Forms Foundations/Continuous Footings
	03	102 Forms Foundation/Isolated Footings
	03	103 Forms Foundations/Piers
	03	104 Forms Foundations/Walls
	03	105 Forms Foundations/Pits, Trenches
	03	106 Forms Foundations/Edge Forms
	03	107 Forms Foundations/Site Items
	03	108 Forms Foundations/Misc.
	03	120 Foundation Structural/Columns
	03	121 Forms Structural/Beam Bottoms & Sides
	03	122 Forms Structural/Supported Slabs
	03	124 Forms Structural/Wall Above First Floor
	03	125 Forms Structural/Stairs
	03	130 Set Column Bolts
	03	131 Set Embedded Misc. Iron Items
	03	133 Set Water Stop
	03	135 Set Column Base Plates

03300		CAST IN PLACE CONCRETE

	03	301 Concrete Foundation/Footings
	03	302 Concrete Foundations/Piers
	03	303 Concrete Foundation/Walls
	03	304 Concrete Pits & Trenches
	03	305 Concrete Slab on Grade ·
	03	306 Concrete Site Items
	03	320 Concrete Structural Columns
	03	321 Concrete Structural Beams & Slabs
	03	322 Concrete Structural Walls
	03	323 Concrete Structural Stairs
	03	324 Concrete Structural Misc. Curbs & Pads
	03	325 Concrete Structural Stair Fill
	03	330 Cement Finish/Float
	03	331 Cement Finish/Broom

03	332 Cement Finish/Trowel
03	333 Cement Finish/Cure & Protect
03	334 Cement Finish/Stairs
03	335 Cement Finish/Patch & Rub
03	337 Grout HM Door Frames

04000   MASONRY

04	200 Masonry Exterior/Face Brick
04	201 Masonry Exterior/Backup Brick
04	202 Masonry Exterior/Backup Block
04	203 Masonry Exterior/Glass Block
04	204 Masonry Exterior/Stone
04	205 Masonry Exterior/Precast
04	206 Masonry Exterior/Paving Brick
04	207 Masonry Exterior/Scaffolding
04	208 Masonry Exterior/Clean & Point
04	220 Masonry Interior/Block 4"-6"
04	221 Masonry Interior/Block 8"-12"
04	223 Masonry Interior/Brick
04	224 Masonry Interior/Stone
04	225 Masonry Interior/Scaffolding
04	226 Masonry Interior/Clean & Point

06100   ROUGH CARPENTRY

06200   FINISH CARPENTRY

06	201 Misc Furring & Blocking
06	203 Wood Studs
06	206 Set Finish Hardware
06	207 Set Millwork

07000   INSULATION

08100   SET HOLLOW METAL FRAMES & DOORS

08200   SET WOOD DOORS

08300   SET SPECIAL DOORS

08600   SET WOOD WINDOWS

The more complex you make a system, the more the chance of its being accurate diminishes. Tailor the system to your particular needs for evaluating costs. A simple system kept up to date is better than a complex system in which reports are delayed and received too late to be of value. A current, simple system allows you to take corrective action when a cost over run situation starts to develop. This is more useful than a lot of late detailed data on why a project has become a disaster. You must be able to make corrections before the item gets so far along it becomes too late to make adjustments. It cannot be said too often: "you must know your costs to properly price new work you are bidding." As you progress in business you will be able to determine how complex a system should be for your needs.

7.5 FIELD RECORDS

Computation of cost figures can be done in the field or the home office. Both systems start in the field with each foreman's daily report form. The sample is shown as Fig. 7.5 and lists the foreman's name, the names of his crew, the total hours worked that day, and the hours each man spent on different types of work.

At the end of each day these forms are turned over to the timekeeper or engineer who performs these functions, depending on how you man the project. First, he checks total hours for each man against his tally. Second, he totals the hours spent on each type work that day by all men on the job. Labor cost for each item of work can then be computed on a daily basis if you so desire. The original is forwarded to the office where costs are kept, and the copy is kept in the job site files.

Unit labor cost for work in place is determined once a week, or more often. A quantity for each item of work put in place is provided by the field engineer. Labor cost spent for each item is divided by the quantity put in place to determine unit labor cost for the period covered.

On some projects all cost computations are made at the home office. The time-keeper simply checks the hours worked and the distribution, for correctness and the engineer provides quantities put in place. These tallies can be incorporated in the payroll form submitted from the field, which can simplify the process.

7.6 UNIT LABOR COSTS

Actual costs are compared with your estimates on a labor statement form, Fig. 7.6. When costs start to go out of line, it is readily noticed on this form. It is important that this system be kept up to date so that corrections can be instigated early on, not when it is too late.

7.6.1 Estimated Costs

Your estimated quantity and costs remain constant unless a change order increases or decreases a quantity and a corresponding dollar change. Column 1 shows the distribution code used for each cost item you are tracking on the project. Column 2 gives a description of the item, and column 3 shows the unit of measure for the item. Column 4 shows number of units estimated, column 5 total labor cost, and column 6 the unit cost used in making up the bid.

FOREMAN'S DAILY REPORT

The Volpe Construction Co., Inc.

DISTRIBUTION

Project: _____

Job No. _____

Date: _____

Trade and Foreman: _____

	Code:													

Total Hours

Name	Hours													

IP-10M 4-74

FIGURE 7.5 Foreman's Daily Report Form

7.6.2 Weekly Costs

Column 7 shows the quantity of work units actually placed during the week; column 8 is the money spent for labor during the week; and column 9 shows the unit labor cost achieved for the week. These weekly costs may be high at the start-up of an item because more time has been spent on layout and other preparatory work. Even so,

Dist	Description	U/M	Est Quantity	Est Cost	Est Unit Cost	CW Quantity	CW Cost	CW Unit Cost	JTD Quantity	JTD Cost	JTD Unit Cost	Rem Quantity	Rem %	Rem Cost	Rem %	CI	(Overrun) Savings
02	050 Demolition	LS	.	69,150						14,313				54,837	79.0		
02	204 Fine Grading	SF	700	2,100	3.00							700	100	2,100	100		
	Total 02			71,250						14,313				56,937			
03	050 Misc Cut & Patch	LS		42,300			133			1,699				40,601	95		
03	101 Form Fnd/Cont Ftg	SF	745	2,010	2.70				357	1,741	4.88	388	52	269	13		(1,624)
03	102 Form Fnd/Insul Ftg	SF	5,891	11,782	2.00				9,243	17,148	1.86	-3,352		-5,366			(5,366)
03	104 Form Fnd/Wall	SF	10,235	31,927	3.12				11,587	30,133	2.60	-1,352		1,794	5		(1,721)
03	106 Form Fnd/Edge Form	LF	200	450	2.25		21		80	138	1.73	120	60	312	69		104
03	107 Form Fnd/Site Item	SF	1,488	3,758	2.53					3,496		1,488	100	262	6		
03	108 Form Fnd/Ms Fnd Forms	SF	3,250	12,200	3.75				2,932	10,423	3.55	318	9	1,777	14		648
03	109 Boxouts	LS		2,500						2,054				446	17		
03	120 Fnd Struct/Column	SF	18,760	37,520	2.00	450	1,062	2.36	16,295	26,848	1.65	2,465	13	10,672	28		6,605
03	121 Form Str/Bm Bot/Sides	SF	6,585	55,680	8.46	600	1,979	3.30	7,543	37,816	5.01	-958		17,864	32		13,064
03	122 Forms Str/Support Slabs	SF	88,125	223,971	2.54	12,000	3,918	0.33	88,321	87,075	0.99	-196		136,896	61		136,702
03	124 Forms Str/Wall Abv 1 Fl	SF	7,170	25,000	3.49	173	471	2.72	1,347	3,145	2.33	5,823	81	21,855	87		8,287
03	125 Forms Str/Stairs	EA	74	3,670	49.59					545		74	100	3,125	85		
03	126 Forms Str/Misc Forms	LF	1,070	4,420	4.13		452			2,274		1,070	100	2,146	48		
03	127 Form Edge	LF	1,060	4,611	4.35				1,000	3,428	3.43	60	5	1,183	25		977
03	129 Const Joint	LF	4,825	6,400	1.33				700	1,096	1.57	4,125	85	5,304	82		(1,172)
03	130 Column Bolts	EA	30	600	20.00									600	100		
03	132 Misc Reglets	LF	40	80	2.00							40	100	80	100		
03	133 Water Stop	LF	150	200	1.33							150	100	200	100		
03	136 Misc Plates	EA	4	60	15.00							4	100	60	100		
03	137 Insulation	SF	700	420	0.60							700	100	420	100		
03	188 Beam Side Form	SF	11,750	58,280	4.96	112	921	8.22	4,612	21,345	4.63	7,138	60	36,905	63		3,856
03	301 Conc Fnd/Footings	CY	608	3,828	6.30				662	4,392	6.63	-54		-564		6	(564)
13	302 Conc Fnd/Piers	CY	23	276	12.00				41	479	11.68	-18		-203		7	(203)
	Total																

FIGURE 7.6 Unit Labor Cost Keeping Form

high unit costs at early stages bear watching to ensure that they drop back into line when production increases and a sufficient number of units are put in place.

7.6.3 Cost to Date

Cumulative costs and averaged unit labor cost are shown in columns 10, 11, and 12. A true picture will emerge as the job progresses, and you will begin to see if original estimates for quantities and unit prices were correct. An overrun on dollars means either the item was underpriced or the quantities were underestimated. Conversely, an under-run of dollars can mean an item was overpriced or the quantities were over estimated. However, a large difference either way can hurt you. Overestimating results in higher bids and lost contracts, and underestimating results in lower bids and lost money. When either of the events take place you should look into the estimator's work to find out whether the takeoff, extension, or checking were accurate. If this check shows problems, it is well to know so you can take corrective action in your estimating department.

7.6.4 Balance to Complete

Quantity of work remaining to be installed is shown in column 13 and the percentage this represents in column 14. Remaining money available to finish the work is shown in column 15 and the percentage to complete in column 16. When work is complete, column 15 shows a gain or loss compared to the original estimate in column 5. On a properly estimated and well managed project, the percent of work remaining in column 14 and the percent of money remaining in column 16 should be about zero. When an item is completed, it should be so noted in column 17. The reason for this is that if you have a quantity remaining to be done in column 13, you can now tell if the estimated quantity was properly taken off or if extra work was performed. This also pertains to column 15 if there are monies available when the item is complete. If the quantity to complete is zero and there are funds available then the job produced better costs than estimated or the item was overpriced.

Before work is complete, projections can be made to see whether you will overrun, underrun, or come out right on the estimate. Column 18 shows this feature. This can be done for both unit price items and lump sum items and corrective action can be taken where needed. The computer is programmed to project, based on costs being achieved, what the final result will be, based on the number of units left to be installed.

7.6.5 Keeping Costs for the Total Project

Costs for the total project can be set up in many different ways. They can be simple by taking your estimate and setting up a cost control sheet, Fig. 7.6.5, which shows in column 1 the code number for the item; column 2 lists the description of the item. Column 3 lists the amount carried on your estimate for the item. Column 4 shows the amount the item was awarded for or the amount carried in the estimate for the items

	COST CONTROL SHEET					DATE 10/14/88						
PROJECT	A A C PROJECT											
JOB # 378		EST	ACTUAL	APPROVED	REVISED	ESTIMATED AMOUNT CARRIED			ACTUAL EXPENDITURES TO DATE			
CODE	LINE ITEM	AMOUNT	AMOUNT	CHANGES	TOTAL	MATERIALS	LABOR	INS/TAX	MATERIALS	LABOR	INS/TAX	TOTAL
	WORK WITH YOUR FORCES											
	GENERAL CONDITIONS	$200,000	$200,000	$7,500	$207,500	54,000	112,000	41,500	25,000	56,000	20,000	101,000
	CONCRETE	$750,000	$750,000		$750,000	300,000	350,000	100,000	250,000	280,000	80,000	610,000
	MILLWORK INSTALL	$48,000	$48,000		$48,000	5,000	31,000	12,000	0	0	0	0
	INSTALL HM / HARDW / DRS	$36,000	$36,000		$36,000	1,000	25,500	9,500	0	0	0	0
	SUB-TOTAL	$1,034,000	$1,034,000	$7,500	$1,041,500							711,000
	SUBCONTRACTORS											
02	EXC. & SITE WORK	$120,000	$122,000	$28,000	$150,000							125,000
02	LANDSCAPING	$25,000	$25,000		$25,000							0
04	MASONRY	$126,000	$124,000	$4,000	$128,000							75,000
05	STRUCTURAL STEEL	$458,000	$453,000		$453,000							350,000
05	MISCELLANEOUS IRON	$59,000	$59,000		$59,000							25,000
06	FINISH CARP/MILLWK MAT.	$20,000	$21,000		$21,000							0
07	ROOFING & SKYLIGHTS	$78,000	$76,000		$76,000							4,000
07	WPFG / DMPG	$10,000	$10,000		$10,000							0
08	HOLLOW METAL MAT.	$15,000	$14,500		$14,500							0
08	HARDWARE MAT.	$21,000	$21,000		$21,000							0
08	WINDOWS/GLASS/GLAZING	$76,000	$72,000		$72,000							0
09	DRYWALL	$117,000	$120,000	$12,450	$132,450							0
09	PAINTING & GLAZED WALL	$47,000	$44,000		$44,000							0
09	ACOUSTIC CEILINGS	$21,000	$20,000		$20,000							0
09	CERAMIC TILE	$1,200	$1,200		$1,200							0
09	FINISH FLOORING	$34,000	$34,500		$34,500							0
10	TOILET PARTITIONS	$1,000	$1,000		$1,000							0
10	TOILET ACCESSORIES	$700	$700		$700							0
15	PLUMBING	$38,000	$42,000	$14,500	$56,500							10,000
15	FIRE PROTECTION	$56,000	$52,000		$52,000							5,000
15	HVAC	$234,000	$230,000	$4,300	$234,300							30,000
16	ELECTRICAL	$186,000	$185,000	($2,800)	$182,200							20,000
					$0							
	SUB-TOTAL	$2,777,900	$2,761,900	$67,950	$2,829,850							1,355,000
	FEE	$150,000	$150,000	$3,500	$153,500							
	TOTAL	$2,927,900	$2,911,900	$71,450	$2,983,350							
	DIFFERENTIAL		$16,000									
	TOTAL AVAILABLE											

FIGURE 7.6.5 Total Project Cost Keeping Format

BALANCE REMAINING THIS DATE				ESTIMATED COST TO FINISH			TOTAL TO FINISH	BALANCE AVAIL	ADD PENDING CHANGES	PLUS OR MINUS
MAT	LAB	INS/TAX	TOTAL	MAT	LAB	INS/TAX	FINISH	AVAIL	CHANGES	MINUS
29,000	56,000	21,500	106,500	29,000	50,000	19,500	98,500	8,000	500	8,500
50,000	70,000	20,000	140,000	52,000	60,000	18,000	130,000	10,000	8,500	18,500
0	0	0	0	0	0	0	0	0	0	0
0	0	0	0	0	0	0	0	0	0	0
			246,500				228,500			27,000
			25,000				25,000	0	0	0
			25,000				25,000	0	0	0
			53,000				53,000	0	0	0
			103,000				103,000	0	0	0
			34,000				34,000	0	0	0
			21,000				21,000	0	0	0
			72,000				72,000	0	0	0
			10,000				10,000	0	0	0
			14,500				14,500	0	0	0
			21,000				21,000	0	0	0
			72,000				72,000	0	0	0
			132,450				132,450	0	0	0
			44,000				44,000	0	0	0
			20,000				20,000	0	0	0
			1,200				1,200	0	0	0
			34,500				34,500	0	0	0
			1,000				1,000	0	0	0
			700				700	0	0	0
			46,500				46,500	0	0	0
			47,000				47,000	0	0	0
			204,300				204,300	0	0	0
			162,200				162,200	0	0	0
			1,390,850				1,372,850			27,000
										153,500
										180,500
										16,000
										196,500

FIGURE 7.6.5 (cont.)

you will do with your forces. Notice that items under your direct control such as general conditions should be shown at the top of the report for ease of reviewing. These amounts may vary because in awarding subcontracts, you clarify what is included in each sub-bid. You may take out some item which may be done with your own forces or add some item of work that may be to your advantage for the subcontractor to do. The cost of bonds can also change the figures. Column 5 shows any change order amount added or deducted to each item of work. Column 6 shows the revised total for each line item. For the work you will be doing with your forces, the balance of the columns will come into play. Columns 7, 8, and 9 are your estimated amounts broken down into labor, materials, and insurance and taxes. Columns 10, 11, 12, and 13 show the actual amount spent to date. Columns 14, 15, 16, and 17 show the balance remaining in each line item. Columns 18, 19, 20, and 21 are a projection of what it will cost to finish each item. This amount should not just be the differential between what was carried and what is left. If that is what you do then forget the exercise because nothing will be gained and you will have to wait until the end of the job to find out the results. Column 22 shows the amount of money that the item will run ahead or behind. During the progress of the job certain change orders will be pending waiting for a resolution or signatures. These should not be entered into the system until they are made final because they may not resolve as expected. This will in some cases show a greater cost, which can throw your numbers out of proportion, and for this reason we add column 23, where you can add the expected dollar value for pending change orders. Column 24 now shows the expected overrun or underrun for all items with the final total for the project.

7.7 WHERE TO KEEP COSTS

Should the cost system be operated at the job or the home office? On a large project with a full office setup, costs can be kept in the field. On small projects, data will be gathered in the field and sent to the home office. On a medium sized job it could be done either way.

Where you have a mixture of large, medium, and small projects, both systems can be used to advantage. Certainly the home office is better equipped to handle volumes of cost data particularly with a good computer system. Whatever the choice, the important thing is that proper records *must* be kept.

7.8 BENEFITS FROM AN EFFECTIVE SYSTEM

An effective cost system provides data early enough to take corrective action before the work on any item is too far along. Bad situations can be avoided. Accuracy of your quantity takeoff is readily checked and this helps produce better estimates on future work. Good estimates based on accurate background information are especially useful when negotiating new work from preliminary plans so an accurate price can be arrived at. Developed costs are needed for guaranteed budgets. A good knowledge of costs helps your relationship with owners and architects when they are planning a project.

7.9 PURCHASING

Purchasing is handled in several ways. Subcontracts in our estimation should be awarded by the estimating department. They may also deal directly with suppliers on other large quantity items such as concrete, brick, and block.

Smaller items such as forms accessories, anchors, and lumber are usually purchased by the project manager or the superintendent. They know the daily needs of the job and handle this phase in stride. Small tools and rough hardware purchases can also be handled by the superintendent, engineer, or whoever is assigned the responsibility in the field.

On a large project a full-time purchasing agent handles all of the details of purchasing. With or without a purchasing agent, purchase order forms provide control for expenditures and save time in checking bills.

The sample purchase order form, Fig. 7.9 has five parts. The top copy (noted Job Copy) is filled out by the person ordering the material. All items are completed,

SUPPLIER'S COPY

PURCHASE ORDER
THE VOLPE CONSTRUCTION CO., INC.

ORDER NO. 28700   JOB NO.   DATE

TO: Ship to: THE VOLPE CONSTRUCTION CO., INC.
                                       Project:
                                       Address:

                                       Via:
                                       You agree to make delivery

Invoice to: **THE VOLPE CONSTRUCTION CO., INC.**   Order Number must appear on all Shipping Papers, Bills,
             54 Eastern Ave.   Correspondence, and exterior of all packages.
             Malden, Mass. 02148
     Send Invoice in **DUPLICATE** and address all correspondence   SHIPPING TICKETS MUST ACCOMPANY ALL SHIPMENTS
                     as above.

Price . . . . Distribution Code . . . . Ordered By . . . .
Terms . . . . F.O.B. . . . . Received . . . .
Confirmation of our Mutual . . . . Agreement With . . . .
Voucher Invoice #/Delivery Slip # . . . . Job Approval . . . .

Please furnish the following:

1.   Supplier agrees to comply with the provisions of the Occupational Safety and Health Act of 1970 and the standards and regulations issued thereunder and certifies that all items furnished under this
     Order will conform to and comply with said standards and regulations. Supplier further agrees to indemnify and hold harmless Buyer for all damages assessed against Buyer as a result of Supplier's
     failure to comply with the Act and the standards issued thereunder and for the failure of the items furnished under this Order to so comply.
2.   Vendor shall be responsible for any labor, material costs and consequential damages incurred by vendee as a result of defective material and delivery delays.
3.   *Right To Know* — Each Supplier is required to implement the provisions of Chapter 470 of the Acts of 1983 of the General Laws of Massachusetts, the so-called "Right To Know" law, upon its
     effective date. When furnishing any material listed on the Massachusetts Substance List, the Supplier will furnish the required Material Safety Data Sheet, together with appropriate labels and
     employee training or instruction material for that substance.

ORDER ACKNOWLEDGED AND ACCEPTED	THE VOLPE CONSTRUCTION CO., INC.
Firm	
By	By _____
Date	
Tax I.D. #	We reserve the right to cancel this order without charges or pen-
Please return signed blue copy to: THE VOLPE CONSTRUCTION CO., INC.	alties if the above delivery date is not met.

FIGURE 7.9 Purchase Order Form

particularly price and terms. The remaining four copies are typed and distributed as follows: copy 2 to the home office, copies 3 and 4 to the supplier, who signs and returns copy 4 as acknowledgment of receipt and acceptance of the order and its terms. Copy 5 remains at the job along with the original copy 1. It is advisable to color code the copies for easy identification.

When material is delivered to the job, copy 5 is marked "received," and is dated and forwarded to the home office so they know:

1. Material has been received
2. Bill can be paid when due
3. Price is correct .
4. Terms of discount (if any)

7.10 CHANGE ORDERS

As stated earlier there is not a sizable project built without change orders. It is essential to have a good workable system for handling all necessary paper work, with responsibility for each action assigned to a specific individual or department.

If a change order is necessary, a quick answer is important. If an answer is not readily available, delays can and do occur. A change is generally started in the field, although it can be initiated by the owner or the architect to increase areas, change layout, install a new piece of equipment not previously contemplated, or for any number of reasons. A change is requested and the first document prepared is a change estimate. When the change estimate is approved, it is then incorporated into a change order. Change orders can be either adds or deducts depending on what is involved.

7.10.1 Change Estimates

The best way to handle change orders is when the change first materializes—to give it a number such as C.E. 1 (for change estimate number one). A good form for submitting change estimates is shown in Fig. 7.10.1. This shows the description of the change and the breakdown of the total monies. You list work by your forces and by the subcontractors. These should be broken down with sufficient detail to make it clear enough so it can be easily understood by the owner and the architect. If you do not have sufficient detail, time is lost because a breakdown will be requested and time passes waiting for you to submit it. It is important to get this information in rapidly.

7.10.2 Change Order Recap

A change estimate is not an order to proceed, but a clear system of showing the information utilized to arrive at the amount of the change. The owner, architect or engineer will issue a change order utilizing his own numbering system. He may combine several of your change estimates into one change order. To keep track of all

THE VOLPE CONSTRUCTION CO., INC.

54 Eastern Ave., Malden, Mass. 02148   Tel: (617) 322-8430

CHANGE ESTIMATE NO. V-

To: **Date:**

                                                                                 Add $

                                                                                 Deduct $

Project:

We submit for your approval ———— copies of Change Estimate V- ——————————— in the amount of $

DESCRIPTION OF CHANGE:

DETAIL OF COST:

Copies of back-up material attached————   Extension of time required:————work days.

                                                          THE VOLPE CONSTRUCTION CO., INC.

                                             By——————————————————————

3M-5-86-IP

FIGURE 7.10.1 Change Estimate Form

THE VOLPE CONSTRUCTION CO., INC.

PROJECT NO. _____

CHANGE ORDERS

PROJECT_____

		DESCRIPTION	CHANGE ESTIMATE		CHANGE ORDER		SUBCONTRACTORS		POSTED	
	VOLPE C.E. No.		DATE	AMOUNT	C.O. No.	AMOUNT	NAME	AMOUNT	SUB	BKPG

IP-500-6-83

FIGURE 7.10.2 Change Order Recap Form

change estimates and change orders, a recap form is shown in Fig. 7.10.2 and is used as follows:

Column 1   Change estimate number or numbers (C.E. #)
Column 2   Description of change
Column 3   Date submitted
Column 4   Amount of the change estimate
Column 5   Change order number from the owner

Column 6   Amount (may include several change estimates (that have been added together)
Column 7   Subcontractors (if any) and amount
Column 8   Posting to subcontract folder and accounting department

When you set up your change estimates file it is generally by number and when they are incorporated into a change order it is very easy to track all the change estimates utilizing this form.

7.10.3 Change Order Cost Distribution Breakdown

Your cost system and monthly requisitions are affected by change orders. The change order cost breakdown form, Fig. 7.10.3, enables your accounting department to keep up to date on all job cost figures which are affected by the change orders.

Cross reference is simplified by showing the owner's change order number and your change estimate number. Amount of increase or decrease is easy to see along with the codes and breakdown of cost items in the change. You will note on this form such items that are possibly done with your own forces such as excavation, concrete, masonry and carpentry items. The subcontractors that are affected are also listed. This form makes it easier and clearer for your accounting department to follow and record. You know at all times the status of every change estimate. By keeping changes in a logical manner, you are able to update subcontracts, owner contracts, and requisitions. This format can be altered to suit any particular project you work on, thus making it easier to deal with changes.

7.10.4 Change Order Authorization

Written authorization to subcontractors or material suppliers affected by owner-approved change orders is essential to keep the process under control. The amount added or deducted is listed along with change order number and change estimate number on the form shown in Fig. 7.10.4. Included is the description of the work encompassed by this change order. It is important to reference all these forms to both the change estimate number and the change order number. In this way you avoid confusion and worry about the status of any change or whether it has been posted.

7.11 CONCLUSION

A good cost-accounting system is the heart of a construction firm, and its importance cannot be overemphasized. Costs and quantities for each project must be compared to the original estimate as work progresses so that you will know how the financial condition of each project will come out.

Purchasing done in the field must be controlled by a system that reports prices, terms, and delivery so that bills can be properly paid and charged to the correct line item by the home office.

THE VOLPE CONSTRUCTION CO., INC.

Project_____ No. _____   Owners C/O No._____

Date_____

Original Contract Amount	$
Previous Change Orders	$
Amount of this Change Increase	$
Amount of this Change Decrease	$ ( )
Contract Amount through this Change	$

C.E. No._____

BREAKDOWN

	Materials		Labor		Code	Quan.	Labor	
Overhead	$		$					
Ins. Tax's								
Excavation								
Fo·e s F.f.								
Forms Struct. F.S.								
Concrete Foot. Cf.								
Concrete Struct. Cs.								
Cement Finish CX								
Masonry Ext. Me								
Masonry Int. Mi								
Rough Carp. RC								
Finish Carp. FC.								
Imbedded Items II								
Material Total	$				Labor Total			
Labor Total	$							
Sub Contracts Awarded	$							
Profit	$					G.L. S.L.		
Total Change Order	$							

	Dr.	Cr.		
Accts. Rec.	Dr.	Cr.		
Cont. Liab.	Cr.	Dr.		

Sub - Contractors

	$	
_____		
_____		
_____		
_____		
_____		
_____		
_____		
_____		
Total Sub - Contracts	$	

Subs Payable Ledger   [   ]

Subs Folder   [   ]

G.L.   S.L.

	Dr.	Cr.		
Subs Awarded	Dr.	Cr.		
Subs Payable	Cr.	Dr.		

Date Posted	by

FIGURE 7.10.3 Change Order Cost Distribution Form

Cost information is a management tool for successful operation. Without this data, you will have a hit-or-miss operation and never know until too late whether money is being made or lost on each of your projects.

As mentioned in various parts of this book computers now play a vital role in all of the actions we take. Accounting, cost keeping, estimating, statistical data, personnel

CHANGE ORDER

THE VOLPE CONSTRUCTION CO., INC.
54 Eastern Avenue, Malden, Massachusetts 02148 • (617) 322-8430

PROJECT:

DATE:

ARCHITECT'S CHANGE ORDER NO.

VOLPE CHANGE ESTIMATE NO.

JOB #

SUBCONTRACTOR'S CHANGE ORDER NO.

TO:

ORIGINAL CONTRACT AMOUNT $

PREVIOUS CHANGE ORDERS $

THIS CHANGE ORDER AMOUNT $

REVISED CONTRACT AMOUNT $

In accordance with the terms of our Contract Agreement covering the above named project, the following changes are hereby authorized:

All other terms and conditions of the Contract Agreement, as they may heretofore have been modified, shall be and do remain the same.

CC:

THE VOLPE CONSTRUCTION CO., INC.

BY _____

TITLE _____

FIGURE 7.10.4 Change Order Authorization Form

records and almost everything that requires record keeping can all be accomplished on a computer. There are many programs available today that were not available five years ago. These programs are upgraded by the software companies periodically to make them more and more powerful. Systems needed to meet your requirements can be programmed to your specifications by any good software company.

Construction association exhibits always have displays by many manufacturers of computer hardware and software developers that are not only informative but give you ideas so you can decide the direction you should be heading with the programs most beneficial to your needs.

Computer hardware is another area where some people get confused. A safe route to take is to select the software package you feel you want and then the hardware. Make sure the hardware can utilize the programs you buy to the fullest extent possible. Computers never seem to have enough capacity after a short period of time because you find you are loading more into them. (Some programs require a lot of memory.) Therefore, the hardware you purchase must have expansion capabilities. You must also be able to network the equipment in the event you set up computers in other areas of your office or on the job site. This feature may cost more at the outset but saves a lot of money when future expansion is needed.

CHAPTER 8

Job Organization

8.1 IMPORTANCE OF EFFECTIVE SUPERVISION

Skillful supervision is essential for bringing each project to a successful completion. Without effective control and direction, everyone goes about his own merry way, which can cause chaos. Problems build on problems, the project is not completed on time, and there are serious cost overruns. The result can be catastrophic and sometimes damage your reputation.

8.1.1 Job Organization Chart

An organization chart for a large project is shown in Fig. 8.1.1. Job functions are applicable to any size project. On a smaller project, one person may fill several slots.

Note that the general superintendent and project manager are bracketed together. They must each be aware of all that is going on. Even though their functions are different, their roles are intertwined and necessary for the successful conclusion of a project.

On smaller jobs, accounting and payrolls may be done in the home office. Even the project manager may work out of the home office because he can handle more than one small job. The general superintendent is not necessary unless you have a number of complex projects. A general foreman can be omitted on smaller jobs where the superintendent can deal directly with the individual foreman for each trade.

8.1.2 Supervision of Subcontractors

Subcontractors generally do trade specialty work on every project, large or small. Today more and more items of work are sublet. In fact, there are some contractors

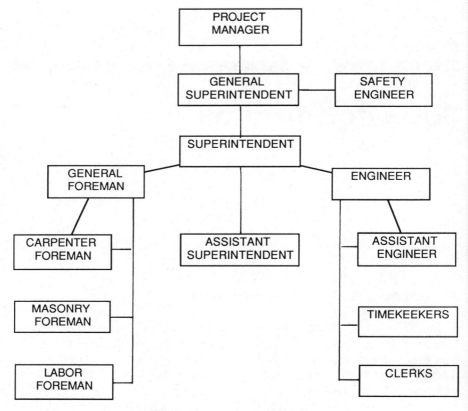

FIGURE 8.1.1 Job Organization Chart

who sublet everything and only provide the top supervision. These contractors are known as "brokers."

Speed of the job can be best controlled where some of the key work is done with your own forces. For instance, a reinforced concrete frame building is difficult to schedule when work is done by a so-so subcontractor. When the work is done with your own forces it is much easier to control the schedule. If needed you can work selected overtime to bring the work back to the scheduled time if it starts to run behind time.

As soon as you are awarded a general contract, a refined progress chart should be prepared. The subcontractors should be informed of your planned schedule. As subcontracts are being awarded, you should review the schedule with all the sub-contractors you are talking to. His ability to meet the time schedule should be weighed just as carefully as the price factor. Proper selection of subcontractors will make project supervision an easier task. Once all subcontractors are awarded and the time frames agreed to by them, the progress schedule is made final and distributed to all parties involved.

Advance planning prevents problems. The subcontractor must know when shop drawings or material samples are to be delivered for approval. Equal employment opportunity requirements, if any, should be discussed. Insurance and bonding are also reviewed and agreed to.

Subcontractors need direction to complete a project expeditiously and economically. The direction must be given by you early on so all the players are headed for the same goal. Without this direction the project will flounder and fail. The single contract system make this goal easier to achieve.

8.1.3 Shop Drawings

Shop drawings are required to show how some elements of the project will be fabricated and also how the item will fit in with other parts of the job. It is vital to know the status of all shop drawings because delays can occur if they are not submitted on time. A good system for keeping track of shop drawings from subcontractors, your review and the architects review is shown in Fig. 8.1.3.

This shop drawing record form provides the following data:

Column 1	Print number
Column 2	Subject
Column 3 & 4	Number of copies and dates received
Column 5 & 6	Date of the first submission the number of copies and the date sent to the architect for approval
Column 7, 8 & 9	Action taken showing return date, number of copies, and the symbol code showing the approval or disapproval action
Column 10 & 11	Second submission, if needed
Column 12, 13, & 14	Second set of action if needed
Column 15, 16, 17, 18	Disposition of prints including date, and the distribution of prints
Column 19	Spare
Column 20	Remarks

This form, as noted at the bottom is made up for each subcontractor so his record is easy to check. It also shows the symbols used for the type of approval given in the sym (symbol) column. This form is essential to keep track of shop drawings so you know at all times where the drawing is and what its status is insofar as approval or rejection is concerned. You will also be able to determine if anyone is taking more time than allowed to make up, review, or approve the drawing.

8.1.4 Transmittal Form

Transmittal forms shorten the amount of time taken to submit or forward drawings, samples, etc., to the owner, architect, or subcontractors. The form has many items listed for which it may be used and they may be altered to suit your company's needs.

RECORD OF SHOP PRINTS

THE VOLPE CONSTRUCTION CO., INC.
54 EASTERN AVENUE, MALDEN, MASS. 02148

PROJECT: _____

JOB NO. _____

DRAWINGS

NO	SUBJECT	RECEIVED		1ST SUB		ACTION			2ND SUB		ACTION			DISPOSITION OF PRINTS				REMARKS
		NO	DATE	NO	DATE	NO	DATE	SYM	NO	DATE	NO	DATE	SYM	DATE	OFFICE FILE	JOB FILE	SUB FILE	

SUB CONT'R _____

DIV. OF WORK _____

SYMBOLS
A APPROVED
B APPROVED AS NOTED
C APPROVED AS NOTED RESUBMISSION REQUIRED
D DISAPPROVED

FIGURE 8.1.3 Shop Drawing Record Form

These forms are filled in by the person forwarding the material (usually the project manager). They may be typed or handwritten thereby saving secretaries' time too. These are multi-copy forms which makes distribution easier. Figure 8.1.4 is a fairly standard form used by a lot of companies.

8.1.5 Subcontractor Folder Information

A folder should be set up for each subcontractor, which will include in addition to the standard correspondence an insurance certificate which is obtained from the subcontractor before he starts work at the site. Be sure the coverage he has meets the requirements of the specifications. If he does not have the proper coverage, you may be liable for the difference. If a bond is required, make sure this is in your files along with a contract signed by all parties.

Items required can be listed on the inside cover of each subcontractor file and checked off as received. Serve notice that if any of these are missing at requisition time, the requisition will not be paid. Requisition time is important to you, and every subcontractor should submit his requisition on or before the required date each month. If a subcontractor's requisition is late, make it plain that no payment will be requested for his work for that month. This makes subcontractors submit their requisitions on time and expedites submission of your payment requisitions.

8.1.6 Engineering Needs

Engineering work at the site depends upon the complexity of the project. A simple warehouse requires someone to give lines and grades.

On a hospital project, lines and grades are only minor items. Layout work requires skill and judgment. Partitions, special outlets, chases for piping, and mechanical equipment have to be set in spaces with minimal tolerances, and exactitude is required.

Your engineer must make sure that the latest and correct drawings are at the site and in the hands of all the subcontractors. This ensures that all plans being used have the latest revision to date, particularly if alterations are involved. A single change order can affect several drawings. If someone on the project is working with drawings that have been superseded it will obviously cause something to go wrong and could cause additional time and cost to correct.

After reinforcing steel has been set by your forces or a subcontractor, the placing should be checked. On some jobs, inspection is carried out by an independent engineering firm engaged by the owner. This is an important item, particularly when architectural concrete is involved. When steel is not sufficiently embedded it will rust, and in time the concrete will spall and you will have an unhappy owner and perhaps a claim against you.

Some contracts require you to furnish manpower to handle such items as quality control for the project. On a large project you could have as many as three or more people involved, covering general construction, mechanical, and electrical work.

TRANSMITTAL

THE VOLPE CONSTRUCTION CO., INC.

54 EASTERN AVE., MALDEN, MASS. 02148   TELEPHONE (617) 322-8430

TO: Date:

                                                                 Project:

Gentlemen: We are sending you the following items:

☐ Enclosed   ☐ Under Separate Cover   ☐ By Messenger

☐ Prints   ☐ Samples   ☐ Letter Dated_____
☐ Shop Drawings   ☐ Specifications   ☐ Change Estimate #_____
☐ Manufacturer's Drawings ☐ Addendum   ☐ Change Order #_____
☐ Brochures   ☐ Certificate of Insurance   ☐

Submitted by_____ Returned by_____

☐ For Approval   ☐ (A) - Approved   ☐ For Job Use
☐ For Final Approval   ☐ (B) - Approved as Noted   ☐ For Quotation Due _____
☐ For Review and Comment ☐ (C) - APN - Resubmit   ☐ Resubmit____Copies for Approval
☐ For Information Only   ☐ (D) - Disapproved   ☐ Submit____Copies for Distribution

Copies	Date or No.	Description	Status

                                                 By:_____

1P-2M-8-72

FIGURE 8.1.4 *Transmittal Form*

Field engineers are also required to keep the "as-built drawings" up to date. "As-built drawings" are required on many projects. These drawings show the actual installed location of all piping, valves, ductwork, fancoil units, electrical conduit runs and junction boxes, etc. In other words, as the title implies, the field engineers do the actual layout of all items installed on the job. As the job progresses the engineer and the trade contractors note on a designated set of plans any deviation required by job conditions or change orders. This set of drawings must be worked on constantly because if you wait until the job is finished to make up these drawings, no one can remember all the true details. Some people get transferred to other projects or leave and the "as builts" are useless. Keeping the "as-builts" up to date on the job and noting the changes as the job progresses make it much easier to produce the final product. Usually you are furnished a set of Mylars to put all this information on. These produce good sharp copies and are more resistant to wear from handling and printing.

8.1.7 Project Meeting Minutes

Job meetings should be held regularly with all major subcontractors and representatives of the architect. The owner and structural and mechanical engineers should also be present when needed. A regular date for this meeting should be established at the beginning of the project so each person can arrange his schedule ahead of time. This ensures good attendance, which helps solve many of the problems and conflicts because all parties involved are present. In this way many of the potential problems are solved before they occur.

Coordination, scheduling of work phases, interpretation of plans and specifications are among the many things aired at these job meetings. You will note on the job meeting report that each item is assigned a number. The item is kept on the agenda of every meeting until it is resolved. This system establishes when the item was brought up, who is responsible for the answer, and when the answer is received. When the item has been resolved, it is removed from the job report. If a lot of items do not seem to get resolved by the architect/engineer, you must call this fact to the owner's attention so he can help get the items resolved.

If the project is delayed because of any item, the job meeting report shows who was delinquent and for how long, thereby establishing responsibility. A copy of a simulated job meeting report is shown in Fig. 8.1.7.1. You will note the date and job meeting number as well as the attendees are noted.

The first column heading denotes who is responsible for any action to be taken. The second column shows a number for each item discussed. The first part of the item number is the number of the project meeting at which the subject was first brought up. The second part of the item number is the number assigned to the item. Item 9-65 means it was taken up at the ninth meeting and the number assigned to the item was 65. As can be seen, you can tell at a glance how long the item has been on the agenda.

PROJECT MEETING #53
MELROSE-WAKEFIELD HOSPITAL
ADDITIONS/ALTERATIONS

Date: June 1, 1988, at 10:00 A.M.

Location: Melrose-Wakefield Hospital

FIGURE 8.1.7.1 Project Meeting Minutes (continued on next page)

Attendees:

W. Hegarty	Melrose-Wakefield Hospital (NP)
M. Mazak	Melrose-Wakefield Hospital
B. Wrightson	Melrose-Wakefield Hospital
W. Flynn	Melrose-Wakefield Hospital (NP)
R. Hall	Melrose-Wakefield Hospital (NP)
N. Nordquist	Ferrenz, Taylor, Clark (NP)
M. T. Yu	Ferrenz, Taylor, Clark (NP)
J. Arabia	London, Kantor, Umland (NP)
R. Genova	Severud-Szegezdy
J. E. Brenner	Volpe
G. Williams	Volpe
R. Lagrega	Volpe
F. Frongillo	Volpe
R. F. Howe	Volpe

RESPONSIBILITY	ITEM #	DESCRIPTION (*–DENOTES CRITICAL ITEM)
LKU/FTC	51-237	*Ductwork 001 & 002* A. LKU to provide drawing for revised ductwork for rooms A001 and A002 to remove low duct drops in corridor. B. FTC to review structure on feasibility of cutting beam under Theta room. C. Direction to be provided by 5/27/88. Per Meeting of 5/20/88. E. Per SS wall at Theta room is OK to cut for duct penetration. Per Meeting of 6/1/88.
CM	51-238	*Theta Room* (Complete) A. CM to release materials for Theta Room. CM to start Theta remodeling upon receipt of materials. Per MWH 5/11/88. B. CM to start construction around July 1, 1988. Completion about August 15, 1988.
FTC	**51-239	*Rebar Shop Drawings* (Complete) Rebar shop drawings for garage footings submitted to FTC on 4/8/88. CM requested immediate turn around to prevent delays on the garage construction. Per Meeting 5/11/88.
MWH/CM	51-240	*Administration & Pavilion* (Complete) A. Demolishing of ADM & Pavilion has been delayed from a start of 5/16/88 to 5/23/88. Revised date now June 1, 1988.

FIGURE 8.1.7.1 *(cont.)*

RESPONSIBILITY	ITEM #	DESCRIPTION (*–DENOTES CRITICAL ITEM)

B. Asbestos removal date has been changed from 5/6/88 to 5/13/88. Revised date 5/16/88.

C. Asbestos to be completed by 6/3/88. Per CM Meeting of 6/1/88.

FTC/MWH/CM   51-241   *Second Floor Renovation*

TC

A. Mechanical and Architectural Drawing due 5/9/88. Not yet per meeting of 5/11/88. CM received drawings 5/18/88. Per Meeting of 5/20/88.

CM

B. CM to bid the following items.
 1. ACT
 2. Carpet
 3. VCT
 4. Millwork
 5. Hardware

Prices will be received from trade contractors for other work.

MWH

C. MWH will supply VWC and carpet for the second floor. Bids for these items will be for installation only.

CM

D. VAV boxes, light fixtures and other long lead items will be released without shop drawing approval to prevent delays.

CM

E. CM to develop schedule for the second floor. Per Meeting of 5/20/88.

CM

F. Base is to be included in the flooring bids. Per Meeting of 6/1/88.

TC

G. Door closers in patient rooms to be deleted. Complete.

CM

H. Door frames to be reused. Complete

MWH

I. All carpet to be supplied by MWH. Broadloom is to be installed in solarium. Complete.

CM

J. Solarium floor elevation is to be raised to unknown height to meet second floor elevator. Complete.

TC

K. TC to review and provide additional information on waterproofing over existing quarry tile on roof deck.

TC

L. TC to review and redesign beam to eliminate narrow step at greenhouse door. Also, hand rails are to be added at doorway.

FIGURE 8.1.7.1 *(cont.)*

RESPONSIBILITY	ITEM #	DESCRIPTION (*–DENOTES CRITICAL ITEM)
TC		M. At shower room 240, a copper pan is to be used under the marble flooring. Complete.
MWH		N. MWH to purchase E-61B, E-62B, E-104H, and S-34. Complete.
TC		O. TC to spec faucets for patient room sinks.
TC		P. TC to provide model # for project manual items missing.
TC		Q. All doors on second floor are to be new red oak. Door sizes and door schedule to be provided.
TC		R. TC to provide model # for plumbing fixtures not noted on drawings.
TC		S. TC to review and advise finish requirements of sink units in patient rooms.
TC		T. TC to provide toilet accessories schedule.
FTC/LKU	51-242	*CAT Scan*
		A. MWH required partial completion of the new CAT scan area by the middle of June.
		B. CM requested the revised architectural drawing and the mechanical and electrical drawings at once. Per Meeting of 5/6/88. Not yet per meeting of 5/11/88. Drawing received by CM on 5/18/88.
		C. CAT Scan work to be a separate project per MWH.
		D. CM is to proceed on a time and material basis. Per MWH meeting of 5/20/88.
		E. Per MWH delivery of CAT Scan is scheduled for June 29, 1988.
MWH		F. MWH to provide dark room equipment to CM for installation.
TC		G. TC to provide fixture model #'s to CM for plumbing equipment and medical gas outlets.
FTC/MWH/CM	51-243	*GARAGE*
		A. CM presented cost reduction to MWH on 4/28/88.
		B. Garage plans to be presented to Appeals Board on 5/25/88.
		C. CM to start garage construction in June.
		D. CM requires revised common footing design on J line. This area is on hold for shop drawings until information is received.

FIGURE 8.1.7.1 (cont.)

RESPONSIBILITY	ITEM #	DESCRIPTION (*–DENOTES CRITICAL ITEM)

                   *   E. Revised garage drawings are required as soon as possible to get hard prices on cost reduction and prevent delays in starting.

                                      F. CM is to release rebar shop drawings on common footing.

                                      G. MHW, FTC, and CM to meet on 5/25/88 to review and finalize cost reduction items.

                                      H. Appeals Board to make decision on June 15, 1988. Per MWH Meeting of 6/1/88.

                                      I. FTC to review drawings and provide to CM a list of drawings and revision dates for construction use. Per Meeting of 6/1/88.

LKU	52-443	*Pharmacy* Complete

                                      A. LKU to review motor size and sheave for increase in fan exhaust to pharmacy hood. Per Meeting of 5/20/88.

                                      B. MJF has installed new motor and shive as directed by LKU.

LKU	52-244	*Histology Leak* Complete

                                      A. LKU to review AC-1 for method to eliminate wall leak.

                                      B. LKU directed MJF on revised piping and pans. Per Meeting of 6/1/88.

MWH	52-447	*Emergency Generator*

MWH to advise CM on date of change over on the emergency generator.

NEW BUSINESS:

FTC	**53-449	*Elevator Infill*

CM requires details for elevator infill to pour concrete slab.

The next meeting to be scheduled at a future date.

Robert F. Howe, Project Manager

RFH/pg
Encl.
cc: All Listed

FIGURE 8.1.7.1 (cont.)

8.1.8 Mechanical/Electrical Meeting Minutes

Some complex projects may require a mechanical coordination meeting in addition to the regular weekly job meetings. These meetings generally, in addition to the contractor, include the mechanical engineer and the mechanical contractors. The minutes of a typical mechanical coordination meeting are shown in Fig. 8.1.8. The same system of numbering and assignment of responsibility for each item is as used in the job meetings previously mentioned.

MELROSE-WAKEFIELD HOSPITAL
ADDITIONS/ALTERATIONS
MECHANICAL/ELECTRICAL JOB COORDINATION MEETING #61

DATE: July 6, 1988

Time: 9:00 A.M.

LOCATION: VOLPE CONSTRUCTION JOB FIELD OFFICE

ATTENDEES:

William Flynn	MW Hospital
Barbara Wrighton	MW Hospital
Richard Hall	MW Hospital
Jim Carroll	M Co.
Jim McHardy	MJF
Harry Woodland	MJF
George Riddle	TGG
Kenneth Smith	NEC
Frank Frongillo	Volpe
Bog Howe	Volpe
Bob LaGrega	Volpe

This is a hard hat project and all personnel are required to wear hard hats at all times while on this project. NO EXCEPTIONS.

RESPONSIBILITY	ITEM #	DESCRIPTION (*–DENOTES CRITICAL ITEMS)
	37-198	*Boston Gas Co. Field Visits*
MWH/LKU	********	2. Waiting for State approval of gas requirements. Per 2/3/88 Meeting. Updated at 5/11/88 Meeting Complete Per Meeting 7/6/88.
		4. Per CM a copy of the gas submittal was sent to Boston Gas Company for their review on 5/31. Per 6/1/88 Meeting.
LKU	*****	5. Per CM Boston Gas states that the information sent to the State needs added information, i.e. detailed plan of gas piping from meter to gas

FIGURE 8.1.8 *Mechanical/Electrical Meeting Minutes*

RESPONSIBILITY	ITEM #	DESCRIPTION (*–DENOTES CRITICAL ITEM)

train. LKU must do to get approval. Per 6/8/88
Meeting. Not yet per 7/6/88 Meeting.

**LKU/BOSTON
GAS CO.,**

6. LKU to contact the B.G. Company for further
information to be sent to the State for approval.
Per 6/22/88 Meeting. Not yet per 6/29/88
Meeting.

MWH/LKU

7. CM received letter from Boston Gas stating
their permission to use H.P. Gas. However,
Mass State approval is still required. Per 7/6/88
Meeting. See attached Boston Gas letter dated
6/9/88.

MJF   41-217

5. Shutdown per MJF is complete and Cleaver
Brooks will be done on 6/13/88. Not yet per
6/22/88 Meeting.

6. A State approval and a Factory Mutual
inspection is required to refire boiler. MWH has
contacted State and FM. Per Bill Flynn. Per
meeting of 6/15/88.

7. Per MWH, the approval of the one converted
boiler is waiting for MJF to get approval by the
Insurance Company and the State. Per 7/6/88
meeting.

ALL TRADES   43-235   *Coordination of M/E Work*

1. All M/E trades are directed to start the M/E
Coordination Drawings to the Basement of the
1963 and 1951 Buildings immediately starting
3/7/88. Per 3/2/88 meeting.

MJF   ************** 3. The 1963 Building Coordination Drawings is to
be started 4/15/88. Per 4/13 meeting. Not yet
per 7/6/88 meeting.

45-254   *Conversion of HP to LP Steam System*

LKU/MJF   Complete per 7/6/88 meeting.

********** 1. Per MJF how are PRV assemblies, etc., to be
changed, to be handled? This is CRITICAL for
LKU to answer immediately. Per 3/16/88
meeting. Not yet per 6/22/meeting.
Complete per 7/6/88 meeting.

********** 2. Per LKU written description to be forwarded
to CM by 4/22/88. Per 4/20 meeting. Not yet
per 6/22/88 meeting. Complete per 7/6/88
meeting.

FIGURE 8.1.8 *(cont.)*

RESPONSIBILITY	ITEM #	DESCRIPTION (*–DENOTES CRITICAL ITEM)

3. Per J.A. the description of work is correct and MJF is to proceed with all PRV conversions on a time and material basis, not to exceed the overall price given to CM. Per 6/29/88 meeting.

47-267   *M/E Required Items*

MJF/TGG/NEC

1. CM has forwarded a letter dated 3/17/88 to MJF, TGG and NEC for outstanding items required per Specifications. These items are as follows. Per 3/30/88 meeting.

TGG

5. Record Drawings are required by each Contractor. The Architect will supply you with (through the CM) a complete set of mylar transparencies CB process to incorporate any and all changes per the General and Supplementary Conditions. Spec. Section S-63, Page 5 and 6, Article 17, Paragraphs 17A.1.1, 17A.1.2, 17A.1.3, 17A.1.4 and Section 01340, Page 1 Paragraphs 1.A, 1.B, 1.C. Updated 5/18/88 meeting.

MJF/M.HANLEY

C. *Fire Protection Systems*

****
3. Metal cabinet with 20 spare sprinkler heads of each type and five (5) Special Sprinkler Head Wrenches per Spec. Section 15200-8, Paragraph 22.A, 22.B, 22.C. Not yet per 6/22/88 meeting.

****
4. Chart Diagram of the Fire Protection System as described in Spec. Section 15200-9, Paragraph 23. Not yet per 6/22/88 meeting.

MJF/M.HANLEY

6. Record Drawings are required by each Contractor. The Architect will supply you with (through the Construction Manager) a complete set of milar transparencies CB process to incorporate any and all changes per the General and Supplementary Conditions Spec. Section S-63, Page 5 of 6, Article 17A, Paragraphs 17A.1.1, 17A.1.2, 17A.1.3, 17A.1.4 and Section 01340, Page 1, Paragraph 1.A, 1.B, 1.C. Updated per 5/4/88 meeting.

7. Operation & Maintenance Manuals for all fire protection equipment as required by the Specifications.

FIGURE 8.1.8 (cont.)

RESPONSIBILITY	ITEM #	DESCRIPTION (*–DENOTES CRITICAL ITEM)

Electrical Systems

NEC — 4. Record Drawings are required by each Contractor. The Architect will supply you with (Through the Construction Manager) a complete set of mylar transparencies CB process to incorporate any and 11 changes per the General and Supplementary Conditions Spec. Section S-63, Page 5 and 6, Article 17, Paragraphs 17A.1.1, 17A.12.2, 17A.1.3, 17A.1.4 and Section 01340, page 1, paragraphs 1.A, 1.B, 1.C. Updated per 5/4/88 meeting.

ALL TRADES — The record drawing item in your respective trade is expected to be completed immediately after receipt of the mylar drawings. Per 3/30 meeting. Updated at 5/18/88 meeting.

49-272 — *New Trane Chiller (AC-1 System)*
Complete per 6/8/88 meeting.

**** 3. Water balance to be completed for chilled water. Not yet per 6/1/88 meeting.

MJF * A. Reports due from MJF 6/22/88. Not yet per 7/5/88 meeting.
Complete per 7/6/88 meeting.

MJF ** 4A. Per MJF the sheaves were changed on 6/21/88. This raised the motor amps of 7.2A. The maximum is 8 amps. Waiting for balance report. Per 6/22/88 meeting.
Complete per 7/6/88 meeting.

LKU/MWH — 6. CM requests LKU and MWH to investigate how the existing coils can be cleaned. This is a maintenance item. Per 6/29/88 meeting.

53-284 — *Fiscal Area Basement of 1951 Building*

MWH/LKU ***** 1. Per MWH the ductwork which feeds this fiscal area must be moved. It cannot be in corridor. Per 5/1/88 meeting. LKU must respond immediately. Not yet per 7/6/88 meeting.

2. LKU to check and provide a sketch before JA leaves today. Per 6/29/88 meeting. Not yet per 7/6/88 meeting.

FIGURE 8.1.8 *(cont.)*

RESPONSIBILITY	ITEM #	DESCRIPTION (*–DENOTES CRITICAL ITEM)

Nitrous Oxide Alarms
Complete per 7/6/88 meeting.

NEC/MWH
5. Per MWH, equipment is on site waiting for Ohio Med Representative to review wiring of alarm panels. Per 6/22/88 meeting.
Complete per 7/6/88 meeting

LKU/NEC
5A. The alarms are to be high pressure, low pressure and reserve. Per LKU and MWH. Per 6/29/88 meeting.
Complete per 7/6/88 meeting.
6. Per NEC, to be completed by 7/6/88 meeting.

53-289   *1963 Building*

MJF
1. T-stats and the duct system is not functioning. MJF to respond and correct immediately. Per 5/11/88 meeting.

LKU   ***
2. Per MJF the 1963 Building is not ducted as shown on the Contract Drawings. MJF required direction on how to correct the problem. Per 5/18/88 meeting. Not yet per 5/25/88 meeting.
3. CM directs MJF & Honeywell to meet with Volpe to explain how Honeywell installed their work for controls in the 61 Building Basement. Per 6/1/88 meeting.
4. Meeting was held with CM, MJF & MWH on 6/2/88.

MJF/LKU
A. Waiting for 63 Building coordination drawings. Per 6/8/88 meeting. Not yet per 6/9/88 meeting.

54-295   *2nd Floor Renovations*

ALL TRADES
1. Per CM, MJF, NEC and TGG are requested to order all long lead times (over 6 weeks lead time). Drawings will be given to all trades for immediate pricing. Per 5/18/88 meeting. Not yet on long lead items.

ALL TRADES   *****
2. Per CM all trades are to be submitted a purchase & delivery (P&D) schedule immediately. Per 5/25/88 meeting. Not yet per 6/22/88 meeting.

TGG/MJF   *****
2A. Received P&G schedule on light fixtures per 6/22. Still waiting for MJF & TGG P&D schedules. Per 6/22/88 meeting. Not yet per 7/6/88 meeting.

FIGURE 8.1.8 *(cont.)*

RESPONSIBILITY	ITEM #	DESCRIPTION (*–DENOTES CRITICAL ITEM)
	56-303	*Garage*
MJF		1. CM directs MJF to start the garage sprinkler drawings immediately for submission as shop drawings. Per 6/1/88 meeting.
M. HANLEY/ MJF		2. MJF to contact M. Hanley to check on status of sprinkler drawings. Per 6/8/88 meeting. Not yet per 6/29/88 meeting.
		3. Per LKU, MWH is to give what is on file with the Building Department for a sprinkler system immediately. Per 6/29/88 meeting.
		4. Per MWH, the lower level will be completely sprinklered and piping for future garage levels will be capped at the ceiling of the lower level. Per 7/6/88 meeting. This system is to be a dry system.
	56-304	*Third Floor Renovation Coordination Drawings*
MJF		1. MJF is directed to start the ductwork coordination drawings immediately. Per 6/1/88 meeting. Complete.
		2. MJF to complete the coordination drawings of third floor ductwork by 6/15/88 for 63 Building and 6/22/88 for the 51 Building. Per 6/8/88.
		3. 63 Building coordination drawing is complete. Waiting for 51 Building coordination drawing. Per 6/2/88 meeting. Complete per 7/6/88 meeting.
		4. Per LKU, the 63 Building changes are extensive for all M/E trades. LKU will need one week to make all modifications before the coordination drawing can be completed and submitted for approval. Per 6/29/88 meeting.
		6. Per CM, the latest architectural drawing which changed the toilets is to be used with the M/E coordination drawing to finally coordinate the 63 Building, Third Floor. Per 7/6/88 meeting.
	56-305	*Smoke Hatch*
NEC		1. LKU has directed NEC to install a Derby electric release on the smoke hatch. Per 6/1/88 meeting.
LKU		2. CM requests LKU to forward a catalog cut and MFG on the smoke hatch release as soon as

FIGURE 8.1.8 (cont.)

RESPONSIBILITY	ITEM #	DESCRIPTION (*–DENOTES CRITICAL ITEM)

possible. Per 6/29/88 meeting. Not yet per 7/16/88 meeting.

57-307   *Fire Alarm 51/63 Buildings*

MWH/NEC   Complete per 7/6/88 meeting.
1. Per CM, NEC has been directed to install new fire alarm system wiring only. All final connection and equipment is to be completed by MWH. NEC is to perform this work on a T&M basis with an estimated cost of $5,000 +/– to be paid directly by MWH. Per 6/8/88 meeting.

MWH/NEC   Complete per 7/6/88 meeting.
2. Per NEC, all work to the firm alarm system is complete and a formal invoice is to be sent to NEC. Per 6/29/88 meeting.

58-308   CAT Scan Coordination Drawings

MJF   MJF to produce coordination drawings for the CAT Scan. Per Meeting of 6/15/88.
Per LKU a shop drawing for ductwork is required for shop drawing submittal and for coordination of the electrical and x-ray equipment work to be done. Per 6/29/88 meeting.

MJF/NEC   1. Per CM, MJF is to locate all duct penetrations with the drywall contractor who will provide all framed openings. MJF is not to modify or change any openings in the framing after this is coordinated by those trades. Per 7/6/88 meeting.

LKU/MWH   Complete per 7/6/88 meeting.
1. MWH requests the CAT Scan to not be on emergency power. LKU to review and respond. Per 6/29/88 meeting.

LKU/MWH   Complete per 7/6/88 meeting.
1A. Per LKU the drawings show the power for the CAT Scan is on normal power. This is OK per MWH. Per 6/29/88 meeting.

58-309   DX Unit
Complete per 7/6/88 meeting.

CM   CM to review existing DX unit on 63 Building roof for Contract responsibility. Per meeting of 6/15/88.

FIGURE 8.1.8 (cont.)

RESPONSIBILITY	ITEM #	DESCRIPTION (*–DENOTES CRITICAL ITEM)

Complete per 7/6/88
1. LKU to review and respond. Per 6/29/88 meeting.
Complete per 7/6/88 meeting.
2. Per LKU, this DX Unit is not to be removed. Per 6/29/88 meeting.

59-310   *Air in 4th Floor*

LKU/MWH

1. Per MWH, in Room 401 there is a complaint of lack of air. LKU to check and respond as soon as possible. Per 6/22/88 meeting. Not yet per 7/5/88 meeting.
2. Per MWH, LKU requests a larger diffuser in Room 401. Per MWH, a price is required prior to this work being done. Per 7/6/88 meeting.

59-311   *SARA System*

MWH/NEC

Complete per 7/6/88 meeting.
1. Per MWH they have requested NEC to complete the SARA System into the 2nd Floor delivery rooms. Per 6/22/88.
3. Per NEC, MWH is to advise NEC on time schedule for completion of SARA System in delivery area. Per 7/6/88 meeting.

60-312   *Garage Drains*

Complete per 7/6/88 meeting.

LKU/TGG

1. Per JA, the garage drains have been approved. Shop drawings will be forwarded to CM as soon as possible.

60-313   *Third Floor Electrical Coordination*

LKU/MWH

1. Per MWH, the electrical drawings and furniture layout needs to be coordinated and some changes to the electrical drawings are to be made. LKU to provide these changes. Per 6/29/88 meeting.
2. Per MWH, the electrical changes are to be made by B. Wrightson today. Per 7/6/88 meeting.

60-314   *AC-1 Unit Air*

Complete per 7/6/88 meeting.

MJF/LKU

1. Per MWH, the AC-1 Unit is not delivering the air requirements. LKU requests the balancing

FIGURE 8.1.8 *(cont.)*

RESPONSIBILITY	ITEM #	DESCRIPTION (*–DENOTES CRITICAL ITEM)

contractor to review and respond immediately. Per 6/29/88 meeting.

60-315   *O&M Manuals*

LKU   ********* 1. CM requests LKU to return the O&M Manuals as soon as possible. They have been outstanding for 8 weeks. Per 6/29/88 meeting. Not yet per 7/6/88 meeting.

60-316   *Third Floor Ductwork in Shaft*

1. There is a conflict with ceiling height with the ductwork coming out of the shaft as shown on HVAC-8. LKU to review and respond. Per 6/29/88 meeting.
2. Review was made with LKU, CM, and MJF. Verbal direction was given to MJF. MJF to show this direction on coordination drawing. Per 7/6/88 meeting.

The next meeting will be held on Wednesday, July 13, 1988, @9:00 A.M., at the Volpe Field Office.

RJL/mlp

cc: All Listed

FIGURE 8.1.8 (cont.)

8.1.9 Daily Reports

Daily reports show what item of work was worked on, where it was accomplished, and the number of workers on the project. Reports can be as simple or complex as you want. The following items are necessary:

The date
The weather
Temperature at starting time and again in afternoon
Name of superintendent
Number of employees working in each trade
Areas where work is being done

The weather is important in the event of a future problem developing: someone may say was caused by material freezing or being improperly installed in temperatures that were too low. The report will verify the actual condition as opposed to someone's guess.

Remarks can show special problems, strikes, or unusual events. Names of visitors can be listed. You may want to use the daily report to list materials received or materials needed.

Quantity of concrete poured that day, with a cumulative total to date, is often part of the daily report. Other critical items can be recorded in the same fashion. Subcontractor activity can be shown.

If there is a delay in completion of a project, you can get a pretty good picture from the daily reports. You will probably know the reasons but the reports will back up the information that is necessary such as:

Bad weather?

Insufficient manpower?

Some type of strike?

Delayed material delivery?

Delayed delivery of owner-supplied equipment or materials?

Insufficient information from the architect or engineer?

Answers to these and many other questions can be found in daily reports. If the superintendent were ordered to increase manpower on the project, this report shows when and how much manpower did increase. The report should be *signed* (note emphasis) by the superintendent and mailed to the home office at the end of each day. The signing shows that he has seen the report and agrees with it. His signature becomes the certification if questions arise at some future date.

A sample daily report is shown in Fig. 8.1.9. Trades are listed to make it easier for field people to complete. This also provides uniformity.

The project manager, general superintendent, and others in the chain of command can initial the report to indicate they have seen it.

8.1.10 Documentation

Documentation is very important, especially in todays litigious society. The job meeting minutes, mechanical/electrical meeting minutes, daily reports, daily work vouchers, transmittals, the shop drawing logs, and progress schedules are all important documents when you have to prove a point at some future date. Correspondence must be answered in the normal course of business events. The specifications, general conditions, supplementary general conditions, and special conditions all have relevant times of which either notices have to be given or replies have to be made. These should all be adhered to in the event a problem arises so you can show a clear record of all events that have taken place. If problems start to occur they will, as a rule, continue.

These problems are not necessarily just with an owner, they can be with subcontractors also. Regardless with whom the problem is, the records are just as important. Everything should be recorded or documented at the time. Material that is developed later is not as effective as that which was developed at the time of the occurrence.

DAILY REPORT

THE VOLPE
CONSTRUCTION CO., Inc.

P. Mgr.		Gen. Supt.	
V.P.			

PROJECT: _____ DATE _____

SUPERINTENDENT: _____

DESCRIPTION OF WORK

No. of Days _____ Day _____
Weather _____
Temp. A.M. _____ P.M. _____
Visitors: _____

PERSONNEL

No.	Contractors TRADE	No.	Sub-Contractors TRADE
	Project Manager		Acoustic
	Superintendents		Drywall
	Office Manager		Electric
	Engineers		Elevator
	Engineers' Aides		Excavation
	Timekeepers		Floors, Resilient
	Material Checkers		Glazing
	Clerical		Heating
	Mason Foreman		" Sheet Met.
	Masons		" Controls
	Mason Apprentices		" Covering
	Mason Tenders		Masons
	Cement Finisher		Mason Tenders
	Carp. Foreman		Misc. Iron
	Carpenters		Painting
	Carp. Apprentices		Plumbing
	Labor Foremen		Plumbing Covering
	Laborers		Roofing
	Hoist. Engineers		Sprinklers
	Oilers		Steel, Rein.
	Pump Operators		Steel, Struct.
	Comp. Operators		Waterproofing
	Truck Drivers		Windows
	TOTAL		TOTAL

I.P.

FIGURE 8.1.9 *Daily Report Form*

8.1.11 Daily Work Voucher

Under special circumstances, the daily work voucher shown in Fig. 8.1.11 is used to keep track of labor, materials, and items of overhead. These circumstances include:

1. Time and material work ordered by the owner or architect. (This might be a change order where cost cannot be agreed on.)

DAILY WORK VOUCHER

Date_____19_____

Charge to:_____Contract Name and No._____

Address:_____ _____

The work described below has been performed for you and the cost thereof has been charged to your account.
Sub-Contractor: The amount will be deducted from our next payment to you for this job.
Dealer: Please mail your credit, or check in payment, **The Volpe Construction Co., Inc.**

LABOR					MATERIAL AND EQUIPMENT			
TRADE	DATE	HOURS	RATE	AMOUNT	QUANTITY	ARTICLE	PRICE	AMOUNT
		Total Labor						
H. & W., Taxes and Insurance					Small Tools			
					Total Material and Equipment			
					Sales and Use Tax			
Travel, Board and Room					Total Labor Including Insurance, etc.		*	
					Job Overhead			
					Profit			
	Total Carried Forward*				TOTAL $			

Description of Work:
_____
_____
_____
_____
_____
_____
_____

Approved by: THE VOLPE CONSTRUCTION CO., INC.

_____   _____
Signature of representative of party for whom work is done.

FIGURE 8.1.11 Daily Work Voucher Form

2. Work done for a subcontractor by your own forces.
3. Extra work done by a subcontractor for you.
4. Work on an existing facility beyond the contract scope that cannot be defined in sufficient detail to come up with firm price.

Signatures of both your superintendent and the representative of the organization for whom the work is performed are essential on this form. They acknowledge the work was done and your superintendent verifies that details are correct.

8.2 SUPERVISION BASED ON SIZE AND SCOPE

Supervision for a small and simple project may require only one man, a super-intendent. A large complex project requires an entire staff for daily operation. It is not unusual for a specification to require certain full-time engineers at the site. For instance, on laboratory construction, a mechanical engineer may be essential to coordinate and check installation of equipment and utilities. This can be helpful and may even save money because mechanical and electrical work is not usually a general contractor's superintendent's strong suit.

Architectural concrete needs more supervision than normal because of intricate formwork and careful placing of reinforcing steel and concrete. The steel rods must be covered with the specified thickness of concrete. Special features such as accent strips may be required.

Machine bases in an industrial or power plant must be placed accurately and anchor bolts for them must be set in the exact location. Much detail checking is required to get things right on the mark whether it be the elevation of the pad or the anchor bolt location. Your engineers help with these items by ensuring the accuracy needed is provided.

Even a simple project sometimes requires more staff than you might expect. Projects spread out over a large site, such as an industrial complex or garden apartment housing, are best handled by dividing the project up into areas. Put a man in charge in each area and set up a central office to handle the paper work of scheduling, ordering, payroll, material purchasing and handling. Each section must be individually scheduled. Material deliveries must also be coordinated for the various areas. Type and cost of supervision must be taken into account when figuring these types of projects.

8.3 COMMUNICATIONS

Cost of communications can get out of hand if the phone is used every time there is a minor question or request. If the project really requires constant touch with the home office, consider a direct phone line—it may work out to be more economical. A direct line might also be useful on a large project closer to the home office—it can be an extension phone from your central switchboard.

The mails can be used to handle most requests on a well managed job. Multi-page reports and plans that are not urgently needed can often be sent parcel post, special handling, at considerable savings in postage compared with regular first-class mail or express delivery. FAX machines have become very popular and with good reason. They can transmit documents in a hurry, particularly if you need one that is signed. Many agencies accept FAX-transmitted documents.

Job visits by project managers or general superintendents should be coordinated to avoid unnecessary travel.

8.3.1 Travel Policy

Projects beyond the normal driving mileage established for your area will require some additional considerations. Reimbursement of expenses for your supervisory personnel has to be considered in the estimate. Arrangements should be settled in advance because these expenses can amount to a sizable sum of money.

Before staffing the project with people from your home office, investigate the local area. Is competent supervision available? Are there good engineers and foremen available? What work will you do with your own forces and what will be subcontracted? When skilled craftsmen are in short supply, you may have to send journeymen from your area and this costs quite a lot of money. Investigate all of these factors before submitting your bid so they do not come as a surprise during construction. Costs for travel, board, and room are carried on the overhead sheet shown in Chapter 4. Just list the people to be sent, the time they will be there, the going rate for room, board, and travel, and the number of weeks they will be on the project to come up with the amount of money to allow.

8.4 OVERTIME CONSIDERATIONS

Overtime should be used only for good reason. It is expensive and does not always produce the desired results. For example, in an area where labor is scarce, one firm may decide to go on overtime. What happens?

At the outset, this firm gets more men. After about two weeks, the men get tired and production falls off. Soon production with overtime can be no more productive than a regular work week. Some men will take off a day during the week and work overtime on weekends. Other contractors go on overtime in order to keep men they have or to pirate men from other contractors. This forces all contractors in the area to go on overtime. When this happens, all expected advantages are gone and the only result is higher costs.

Overtime does have its place. It is necessary when you get behind schedule (when a few hours or weeks of overtime will bring the project back on schedule). Weigh this cost against what the time lost will cost you in the long run. Working a couple of hours overtime on a given day, to finish setting forms or reinforcing steel, can make possible a concrete pour the next day. This gains a full day on the schedule. Delivery and erection of structural steel is usually scheduled a few weeks in advance and some overtime may be needed if you have fallen behind schedule or if structural steel is coming in early. You may work some overtime to make sure everything is ready to receive the steel.

Overtime to finish a concrete floor or complete a concrete pour is not always within your control and is not intended to be part of this consideration.

When a push is needed to complete a project or a phase of one, work with the smallest possible crew. You can save money by establishing a realistic schedule that takes into account the man-days needed to complete the work. Overtime may not be the answer for speeding up a project—second or third crews on shift work may be better because the men are paid at the regular hourly rate.

8.5 CONCLUSION

Job meetings scheduled at regular intervals are always worth the time and effort because not only can you head off delays, but if someone is trying to make a mountain out of a molehill the job meeting reports will give the true picture. Systematic control of the subcontractors' drawings, time schedule, and paper work is part of your responsibility as the general contractor.

All facets of job organization including staffing, communications, travel, and justification of overtime require interest and guidance from top management so that your planned profits do not evaporate as your project nears completion.

Labor Relations

9.1 GOOD LABOR RELATIONS

The feelings and attitude of your employees will affect the growth and profitability of your firm. By showing a genuine interest in all employees, you will get a favorable response. People like being part of a well-managed team. It gives them a certain sense of pride to have a good relationship with the firm.

Employees usually show their feeling of goodwill for your firm in dealings with clients. Clients can always sense the employees' attitude in a firm. A client will often tell you that he enjoys working with certain people in your organization. Likewise, he will tell you when someone is uncooperative or difficult to deal with. This is why good employee relations and team effort will help you maintain good working arrangements with clients. When an owner tells you he is happy with one of your employees, make sure you let that person know what was said. Commend him and let him know you appreciate his efforts and that this kind of comment from an owner is good for the company's future growth.

9.1.1 Fairness

Manpower is not a machine that can be abused and, if it breaks down, be easily replaced. It costs money to train people in your methods and standards. It takes time to integrate them into smoothly running crews. Thus there are good sound economic reasons to provide steady work and to treat employees fairly. They are the backbone of your company.

9.1.2 Loyalty

Loyalty is a result of fair treatment in matters large and small. People like people who care—especially people who care about *them*. Hourly employees appreciate your efforts to keep them steadily employed. A little extra care in dealing with your people can make your firm better than average. A good safety program (see Chapter 11) is essential to good employee relations. Explaining the various benefit packages such as health insurance, vacation time periods, and profit sharing, also prevents any future misunderstanding, which can cause problems.

9.2 UNIONS

Some contractors operate with union shops and others operate with non-union or open shops. You should treat unions just as you would treat their individual members—with respect, fairness, and a certain amount of give and take. By doing this, problems are generally kept to a minimum. Personality clashes will occur from time to time. Remember there are two sides to every story. Try to understand the other side before making a snap judgment. Problems do develop on jurisdictional matters, and these are discussed later in this chapter.

Many contractors do not meet their employees' local union leaders until a problem develops. This is bad. It is better to get to know the leaders in a friendly atmosphere so that, when problems do occur, discussions are more likely to be objective.

Local association work is a good way to meet union leaders. Request a spot on the labor committee or some joint apprentice committee. Attend functions where joint meetings take place. After you get to know one another on a first-name basis, the problems seldom get completely out of hand. Solutions are easier to reach once you have some understanding of one another's goals and objectives. It is a good policy to have a pre-construction luncheon with the business agents who will be involved on each new project. Explain to them what the job entails. They appreciate knowing what is going to happen on the project as well as knowing what the manpower requirements may be. This kind of discussion can sometimes alert you to a potential jurisdictional or manpower problem before it happens so you can take the proper steps to alleviate it.

9.2.1 Work Stoppages

Work stoppages can occur at any time over almost anything. Sometimes the cause seems ridiculous. When a strike is called by a craft, first find out why. Make an objective investigation by discussing the dispute with your own people. If you decide that your job management personnel are wrong, admit it and rectify matters at once. Stalling only makes the situation worse.

When you think you are right, point out the facts to the business agent. Any experienced business agent who finds his people were wrong will say so and get the men back to work.

Strikes are always costly. The sooner they are resolved, the better. Workers do not like to lose time. Furthermore, if you are wrong and the strike is prolonged, their attitude could become antagonistic.

A strike can result from a job locker being too small, not cleaned or not heated. When your contract calls for specific items, you will have to provide them sooner or later. Do it early and avoid unnecessary animosity. Though strikes should be avoided where possible, this does not mean you should give in to unjust demands.

9.2.2 Payment to Union Funds

Health and welfare, pension, vacation, and apprentice fund payments are covered by legal contract terms with the various unions. Payments are due on certain dates. Meet these dates, for word about slow payers gets around. In some situations, slow payment could hurt you because the unions have a right to refuse to man your jobs. You also may have to post a bond to the fund involved, which will cost you additional monies.

9.3 JURISDICTIONAL DISPUTES

A jurisdictional dispute occurs when two crafts claim the same item of work. For example, both carpenters and laborers may claim the unloading and distribution of millwork. Each craft announces that unless their men are assigned to do the work, they will strike! So if you give this work to the carpenters, the laborers will strike, and if you give it to the laborers, the carpenters will strike. In spite of "no strike" clauses in your labor contract, this threat occurs time and again.

9.3.1 What to Do

Determine your past assignment for the item of work. Examine area practice as well as the jurisdictions outlined in the work rules of each craft.

Past agreements by various unions are published in a pamphlet called the Green Book. Some of these agreements go back many years. In addition, there are records available of more recent assignments by the National Joint Board for the Settlement of Jurisdictional Disputes.

9.3.2 Help from Washington

When work assignments are made in accordance with past practices and a strike occurs, wire the National Joint Board for the Settlement of Jurisdictional Disputes in Washington, DC. Explain the circumstances in full detail, and request that the board chairman get the striking union back to work and then resolve the problem.

Ordinarily the chairman will wire the international office of the striking union to get the men back to work. Then he will request both the striking and nonstriking international unions to send representatives to settle the dispute. If this fails, the joint board will review the facts of the complaint at their next weekly meeting. This is why it is essential to provide all the facts relating to the work assignment plus a lucid description of the work in question and past practice in your project area.

At the hearing a decision will be made, and unions that have previously so stipulated will have to abide by the decision. At this time not all AFL—CIO international unions in the construction industry are bound by decisions of the board. Some of the larger cities have a local board of their own.

Another agency that is helpful in settling some strikes is the National Labor Relations Board of the federal government. This board also hears cases on secondary boycott strikes and charges of unfair labor practice by either the contractor or the union.

9.3.3 Local Help

The staff of the local chapter of the A.G.C.A. is well equipped to help in labor disputes. Changes are continually being made in the various procedures and labor laws. The association staff stays up to date on these changes and can help you on the procedure and text of wires to Washington, DC. This is an important reason to belong to a nationally affiliated organization. You give them the same information that would have been sent to the joint board and they will help you resolve the problem.

9.4 AREA DIFFERENCES

When bidding work outside the jurisdiction of your own unions, examine the local area work rules that will apply. Work rules for various trades can differ from one area to another. If you operate open shop, working conditions, wages, and other conditions should also be investigated.

Some locals may require standby time for electricians on temporary lights, and others may not. You need to know about variations in standby time for different trades so that your bids will properly reflect these differences. Ignorance can be costly. This is another area where belonging to a good national association helps. Their local chapters will have all the information and will advise you on the conditions you must know and follow.

9.5 CONCLUSION

Successful contracting is never routine. It requires extra effort from all of your employees. By creating the feeling of fairness, goodwill, and teamwork through your own words and actions, employees will take a real interest in helping your firm succeed.

When your work force is unionized, it is still essential to foster good working relationships with the men as well as the shop stewards and union leaders. Prompt, fair settlement of grievances goes a long way in maintaining satisfactory working arrangements.

CHAPTER 10

Equipment

10.1 NEED

A building contractor usually owns less equipment than a highway or heavy construction contractor. The reason for this is that building construction needs vary widely, from one project to another. Building contractors fall into several categories. Those who do their own excavation, concrete, masonry, mechanical or electrical work have a larger investment in equipment than those who sublet one or more of these elements. Frequently today, the general building contractor does his own concrete, masonry, and carpentry and subcontracts everything else. There is even the contractor known as a "broker," who sublets all trade work and therefore has no need to own any equipment.

Highway and heavy contractors need certain types of equipment on a project from the start to almost the end. It is therefore more advantageous to own it outright because the rental costs will exceed the cost of ownership. They normally will also have the staff to maintain this equipment.

Our discussion about equipment will be general. For specifics, see *Construction Equipment Guide*, second edition, by D. A. Day and N. Benjamin.

10.1.1 When to Own or Lease

Should you own or lease equipment? This question must be answered depending on your needs and the length of time it will be used on the project.

The experienced contractor finds that some items of equipment should be owned and others leased. As a general rule, you will want to own whatever equipment is used continually on a day in, day out basis in your operations. You will normally rent or lease special types of equipment.

10.1.2 Owning

Smaller building contractors usually find that at least one truck crane, the size depending on the type of work he does, is a good investment. They own trucks that are in steady use as well as smaller pieces of equipment such as pumps, mortar mixers, concrete vibrators, and compactors. Frontend loaders, bulldozers, forklift trucks, and small road rollers require careful research of costs before purchase; more on this later.

10.1.3 Renting and Leasing

Today just about anything a contractor needs, from small tools to the largest crane, can be rented or leased.

When a large piece of equipment is needed for a single job, leasing may be the most economical answer. For example, a hoist for a high-rise building must have enough speed and capacity to keep up easily with the scheduled work. Such a hoist is an expensive item and may not be, adaptable to other types of projects you expect to work on in the immediate future. It is the heart of a high-rise job, and "getting by" will something smaller or makeshift does not pay. Any breakdown could tie up concrete trucks and cement finishers for several hours, causing additional overtime costs, including temporary lighting, and other unexpected expenses not in your estimate. A good hoist—both material and personnel—is a typical leased item.

Leased equipment is generally in excellent shape and often brand new. Agreements may include full service, parts, and repairs by the leasing companies' trained crew. They understand the contractor's needs and can help avoid costly downtime.

10.1.4 Lease Purchase Options

Specialized new equipment is usually available on a lease—purchase option. The contractor can learn all about the equipment while leasing it. If it proves to be a good money maker, he may complete the purchase.

Some contractors follow a rule-of-thumb plan where purchase is made after payments exceed 51% of cost. This is not always a "best buy."

Our firm rented some special equipment valued at $45,000 for a particular project. It was profitable and our rentals totaled $31,000 at completion. The equipment was offered to us for only $14,000, which seemed a real bargain. However, we had no future work or even bid prospects where this type of equipment could be used. Although we might have been able to resell at a small profit, our staff was busy on more important work, so we turned down the purchase option. Time proved the decision right. We have never had a need for this type of equipment since, and our $14,000 might have been a total loss. Bargains may be expensive!

10.2 How to Figure Cost

Here is a simplified example of figuring the costs for leasing vs. owning:

A crane costs $200,000 and will be used on current projects for six months. Annual out-of-pocket costs without repairs, depreciation, registration or insurance will include:

Interest on $200,000 at 10% for 6 mos.	$10,000
Fuel and lubrication	$6,000
Operator and oiler at $1500/wk	$39,000
Total	$55,000

Repairs of large equipment can be very expensive. A full-time crew may seem unnecessary, but it may prove essential by either your choice or union regulations.

Rental of the same crane with operator and oiler is $1000 per day. For a steady six months, there would be 115 working days based on 15 days of bad weather or holidays.

$$\text{Rental for 115 days @ } \$1000 = \$115,000$$

There is no capital investment, no worry about repairs, and no idle crew in a slow spell. The chance for breakdown is diminished because specialists keep the equipment in good repair.

However if this is the way the costs would come out, then it would appear to be more feasible to buy.

On a short-term basis, single out-of-pocket costs can be figured on any piece of equipment. Where the choice is not a clear one, a good accountant can analyze other related costs, including depreciation, both short- and long-term, so your decision will be the best one. You must, of course, realize you will tie up capital for these larger investments.

Truck leasing is a very competitive market and may provide an attractive way to conserve capital. Even so, owning is most economical when there is full use. For part-time use, the economics must be investigated. There is no set rule to follow.

10.3 EQUIPMENT RECORDS

Even when a contractor rents or leases larger pieces of equipment, he soon finds his investment in smaller equipment, such as power tools and other machinery, is sizable. Though he may not have difficulty in knowing the whereabouts or condition of a truck crane, keeping track of pumps, generators, and power trowels and other small pieces of equipment becomes impossible without good records.

Up-to-date records save the cost of unnecessary purchase of equipment. The equipment manager usually has two sets of cards, one based on equipment category and type, and a second based on projects where the equipment is being used.

10.3.1 Equipment Service and Location Cards

A record is made out for each piece of equipment. All equipment is marked with large painted numbers for easy identification. This saves time when the name plate and serial number are broken off or lost. Our system is set up with the prefix letter V on all equipment. Smaller equipment has lower numbers. Hand power tools use Series 1 and their identification numbers run V1.1, V1.2, V1.3, etc. An example shows the front and back of the card as Fig. 10.3.1.

Notice the front of the card shows the Item, Description, Maker's Name, and Serial Number. It also shows where the item is located and the transfer dates.

On larger pieces of equipment, there is a full record of repairs by date and cost. At a glance this gives a complete picture on the cost of maintaining any piece of equipment. This is helpful in deciding whether to repair or replace when a piece of equipment gets older. When the frequency of repair increases there is more down time and the cost of repair starts to get out of line.

The service and location card provides a quick way to find out if there is adequate equipment or a surplus on a particular project, considering its stage of completion.

Equipment that is not being used is costly. As new projects start, a review of these cards will give a good idea where some of your idle equipment can be obtained, thereby avoiding unnecessary purchasing or rental.

Often a superintendent will say; "I can't spare this. I'll need it later," or "I need a standby." This may be valid sometimes, but should be judged from a real need and profit point of view. You must have control over equipment at all times to avoid costly overbuying or renting. If the job is not too far from your yard, you do not need a standby on the job.

Whenever a piece of equipment is moved from one job to another, the service card, in the yard file, is moved from the job it was on to the one it is sent to. This keeps your records up to date as to what equipment is on each job. Sometimes different colored clip tabs can be used for each job. These tabs are attached to the service card in the file. This keeps all the cards in one file without moving them around. Note that the card has columns showing the location or job, when it was sent, and when it was returned.

10.3.3 Equipment Transfer Forms

A three-part transfer form is used for all equipment transfers. The person initiating the move fills out the form, shown as Fig. 10.3.3, which lists the piece of equipment being moved and indicates condition of the equipment.

The first copy is sent with the equipment, the originator keeps the second copy, and the third copy goes to the equipment manager at the home office or yard. Notice the first column refers to the number you use for your numbering system. When the equipment is returned to the yard, the forms are filled out by the field person in charge of equipment and two copies go with the equipment and one stays at the job. The service cards are then moved back to the main file or the tabs removed depending on which system used by the equipment manager.

Needless to say the card is not put back if repairs are required. When the repairs are done then the card goes back to the file, which signifies the equipment is ready when needed.

THE VOLPE CONSTRUCTION CO., INC.

Item _____   Description _____   Model # _____

Makers Name _____   Serial No. _____

LOCATION	IN	OUT	LOCATION	IN	OUT	LOCATION	IN	OUT

FIGURE 10.3.1 Equipment Service & Location Card (continued on next page)

SERVICE RECORD

Purchased from_____

Guarantee Expires_____ Service By_____ 19___

Original Cost $_____

SENT	RET'D.	SERVICED BY	NATURE OF REPAIRS	INVOICE	COST

FIGURE 10.3.1 (cont.)

THE VOLPE CONSTRUCTION CO., INC.

EQUIPMENT TRANSFER

JOB NO. _____ AT _____ _____ 19 _____

WE HAVE THIS DAY SENT TO _____

VIA _____ THE FOLLOWING EQUIPMENT:

JAVCCO NO.	ITEM	CONDITION

Received by _____   Sent by _____

DO NOT send defective equipment to another job. Check with MALDEN first.

FIGURE 10.3.3 *Equipment Transfer Form*

10.4 MAINTENANCE

Proper equipment maintenance is a key to ownership profitability. Without a system for maintenance, a contractor is asking for trouble. An equipment breakdown results in costly waiting time.

It is important to keep your equipment in good repair to avoid breakdowns and lost time. If equipment is kept painted and clean as well as in good condition people have a better feeling toward the company. Shabby looking equipment gives the wrong impression, such as that the firm is skimping on maintenance, and may give the feeling that you are doing the same on the job.

Even the breakdown of a small item such as a concrete vibrator or power buggy can slow down or stop a concrete pour. This causes a lot of trouble and added costs. This equipment should be checked and maintained between pours. Always have a standby vibrator. The cost can be minimal compared to the cost of a delay.

For all smaller equipment that may cause a time loss, weigh values and determine if standbys or spares are needed at the job site.

Maintenance of large equipment is a problem for smaller firms without a well staffed shop. Sending this equipment out for all maintenance and repair is costly. This factor should be taken into account when considering ownership of larger equipment and these costs added to the equation.

Cost of maintenance can get out of hand if it is not watched. Check the costs of work done by outside shops. When the amount approaches the cost of having a qualified mechanic on your payroll, then it is time to hire one. You have the advantage of quicker service and closer attention to your equipment. Minor repairs at an early time will prevent major repairs later.

10.5 CONCLUSION

Considerations for ownership of equipment are the need, the cost, the term of immediate use, and possible future uses.

With the steady improvements in size and capacity of construction equipment, most contractors now try to avoid being "iron rich" and "dollar poor." Owning of a fleet of obsolete equipment is a quick road to bankruptcy.

I repeat it is important to use your accountants who can provide further details on the cost advantages of ownership vs. various leasing and renting arrangements as applied to each particular business situation.

Ownership is usually most economical, both short-term and long-term, for all standard equipment items in daily use on a project. Rental or leasing can be considered for special-purpose equipment, extra large equipment, or newly designed unproved devices.

Computers can make record keeping much easier. A good data base program can be utilized instead of a handwritten card file system for the equipment records, job files, transfers, repair costs, etc.

CHAPTER 11

Safety and Insurance

11.1 IMPORTANCE OF SAFETY

Safety on a planned program basis is sometimes neglected by top-level construction management for several reasons. Some contractors do not comprehend the human benefits or the goodwill generated by a safety program. Others do not realize the lost time that results when an accident occurs. Lost time costs money. Insurance rates vary and are based on actual costs to the insurance carrier and past performance.

11.1.1 Safety Affects Your Costs

Insurance premiums are based on the so-called "Mutual of Average Rates." Worker's compensation and comprehensive general liability rates are actuarially established by state and interstate rating bureaus to which all insurance companies subscribe. These rates are based on experience supplied to the bureaus by insurance companies for various classifications of employees and types of accident. The insurance companies take into account the premiums received and the losses incurred, setting up reserves for unsettled cases. They then add operating costs, profits, etc., on a bureau-approved formula. If they took in more money than they paid out (including expenses and profit), the rates would decrease. If the monies paid (including expenses and profit) are greater than the amount paid in, the manual rates would go up. This has been the trend! This explanation is an oversimplification, but there are many books on this subject, and each state's laws vary.

In addition to this general accounting for the entire construction industry, the rating bureau also checks each firm doing business in a state. The insurance carrier is continually checking the performance of each policyholder. This record shows whether a firm is a good or poor risk. Depending on this ledger status, a contractor is charged at the manual book rate, assessed an extra charge, or given a credit. This is applicable for worker's compensation insurance, property damage, or public liability policies.

11.1.2 Cost Savings

Our firm has had compensation credits as high as 36%, which means our cost was only 64% of the manual rate for worker's compensation. We have had credits as high as 55% on our liability policies, meaning our cost was only 45% of the manual rate.

Experience rates are fixed annually at the policy renewal date. You should review these and object if you think they are wrong. A conference with your carrier will define what caused the change.

Thus it should be obvious that SAFETY does play a big part in the operating costs of the construction business. It affects direct costs, indirect costs (which could be staggering), and relations with both employees and owners.

11.2 SAFETY PROGRAMS

Safety programs spell out the principles that govern the company's views toward safety. Accident prevention plans spell out how the program is implemented, giving the details and responsibilities of all the parties involved on the project. Why do some firms have good safety records and others a poor experience rating? There is only one reason: firms with good records have them because top management believes in safety. A safety program has to start at the top or it will not be a success.

It is amazing how people react to a safety program. After it has been running for a while, the workers know top management is interested and watching. They react by thinking "safety" as well as practicing "safety." Furthermore, they realize the presence of concern for them as individuals. There is less lost time and men are working with the firm, not fighting rules. Work areas are cleaner and men tend to keep them that way.

A safety program spells out the reasons, expectations, and how all must get involved. The following is a model safety program:

SAFETY PROGRAM

It is the policy of The Volpe Construction Co., Inc., to promote safe conditions and safe practices as one of the prerequisites of a successful project, and we expect everyone to conduct themselves in such a way that this can be accomplished.

An effective safety program is one that is regarded as an employee's program with the support and assistance of management. It must be based on practical, workable ideas combined to stimulate the interest of employees.

The very nature of the safety program and safety work requires the leadership of supervision, and the degree of progress depends entirely on the degree of leadership management contributes to it.

"Accidents don't happen, they are caused." It is necessary to find out what caused an accident and then remove the cause. This requires a continuous effort. It is the same element that provides good work and good practices. The simple fact, or at least a large part of safety, is knowing how to do a job right.

The fundamental reason for accident prevention is the avoidance of human suffering. There is no other reason that takes precedence. No individual interested in safety, in the true sense of the word, can concede that any reason for preventing accidents rates above the welfare of his employees.

The most important phase of safety engineering is dealing with individuals—human beings. Our employees must be shown, must be convinced, and must realize and know at all times that safety work is being conducted for their benefit and welfare.

Our success in preventing this human suffering can be measured in several ways. One, the amount of compensation and medical expenses paid. Two, the time to investigate, report, and follow up accidents.

We can appreciate that a reduction in accidents cannot be attained without the proper organization. Everyone in the company is required to take constructive action in eliminating accidents and the cause of accidents. Safety is the responsibility of everyone.

Safety is an integral part of company operations and must, therefore, receive the persistent, watchful attention given any other part of contracting.

11.2.1 Setting up Your Plan

Many good accident prevention plans are in operation with different approaches, all leading to the same goal—a safer job. The following is a model but this plan should be reviewed by your insurance carrier and your attorney because of the many changes which take place in laws governing safety. State laws vary from state to state, making these reviews mandatory.

ACCIDENT PREVENTION PLAN

The purpose of the plan is to outline positive actions to be taken in preventing accidents which cause personal injuries, property damage and interruption of work.

1. Responsibilities

Management   Responsible for planning deliberate accident prevention measures, providing safe equipment and working conditions, training a competent and safety-minded force, and maintaining records prescribed for accidents, injuries, and illness.

Supervisors   Responsible for training and correction of deficiencies, unsafe conditions, or defective equipment when detected or reported.

Workers   Responsible for use of safety equipment; working with deliberate thought for effects on others of their acts; reporting all unsafe conditions, defective equipment, and injuries immediately to foremen.

Subcontractors   Responsible for full provisions of this plan. *EACH SUBCON-TRACTOR SHALL BE RESPONSIBLE FOR THE FOLLOW-ING:*

a. Providing trained, qualified supervisor and workers.
b. Providing and ensuring that the employee wear the necessary personal protective equipment (A shirt will be worn at all times).
c. Providing for a safety coordinator in writing, with the name of a qualified individual in your organization to contact for emergencies.
d. Providing a safe working area, sound equipment, and good tools for the workmen.
e. Cleaning work areas where the workmen have completed their work and during working hours, maintaining high standards of housekeeping at all times.
f. Providing adequate supervision to ensure compliance with safety rules.
g. Providing all power hand tools with grounding wires if the hand tools are not insulated (shockproof).
h. Holding safety meeting with subcontractors' employees as needed.
i. Contacting loss control engineer of your insurance company and arrange a schedule of inspections for the duration of the job.
j. Cooperating in correcting unsafe conditions around the job site.
k. Instructing employees in the safety requirements and requiring them to follow the safety rules and regulations of the project.
l. Inspecting your work area regularly to correct any safety deficiencies that exist.
m. Attending safety meetings as scheduled by the safety director.

Quality control personnel will incorporate surveillance over all accident provisions. Licensed rigger will supervise all rigging.

2. Training

Each employee will be instructed in the company safety policy, this accident prevention plan, and hazards of the work.

Each employee will be instructed in correct method of lifting and obtaining assistance for heavy or awkward loads, and special instructions will be given in preventing falls, which cause one-third of all accidents.

All personnel will be directed in calling for emergency help. Emergency telephone numbers will be conspicuously posted.

Dr._____ Address_____ Tel. No._____
Hospital_____ Address_____ Tel. No._____
Fire Dept._____ Address_____ Tel. No._____
Ambulance or Rescue Service_____
_____ Address_____ Tel. No._____
Police_____ Address_____ Tel. No._____

This plan, posters, and bulletins will be available to all workers.

Insurance company (not agent) to assist in training.

3. Sanitation

Drinking water will be carried in approved containers and cleaned and refilled daily. Paper cups will be furnished and place provided for disposal.

Portable chemical toilets will be provided near the work sites until permanent toilets are available. Replacement toilet supplies will be furnished, and equipment kept clean and sanitary.

Washing facilities will be provided and a supply of paper towels.

Approved first aid cabinet will be installed in first aid room. All injuries, no matter how slight, will be reported for treatment. First aid will be given at the project only by designated personnel, who are qualified by training through the American Red Cross. When there are at least 300 workers on the site, an infirmary will be manned by a registered nurse-full time. No others will treat any injuries at the project.

Complete record of all injuries and illness will be posted within four days of occurrence. OSHA Form 200 or required form will be kept up to date at the site.

Workers will be instructed in identifying and avoiding or disposing of stinging insects, snakes, or rodents and poison ivy if encountered on the job.

4. Fire Prevention

Fire extinguishers of approved types will be furnished on the project. Hoses and portable fire prevention equipment will be stationed at storage and shop areas, with welding machines, on motorized equipment, and with each potential source of flame.

No trash or debris will be burned on the project. Explosives will not be used unless proper permits are obtained.

"NO SMOKING" signs will be posted and careful watch kept on areas containing flammable liquids.

Flammables will be handled only in approved safety cans. Engines will be shut off while fueling.

Oily rags and waste will be kept in covered metal containers. All trash and waste will be disposed of daily. Tarpaulins will be flameproofed.

Welding gas cylinders will be secured upright, capped when not in actual use, and shielded from direct sunlight. Hoses and gauges will be checked for leaks, and kept away from oil and grease. Flowback and explosion of gases will be prevented by installation of safety check valve on each gas torch.

Fire watch will be kept for one (1) hour after all cutting or welding. Falling sparks from cutting or welding shall be caught in metal or fireproof screens.

Temporary heating units will include only components approved by the Underwriters Laboratories or approved federal, state, or governing agency. Operators are to be instructed in the manufacturers' recommendations for safe use and maintenance of heaters. No unit will be set closer than ten (10) feet to wood, fabric, or other flammable material. Natural or fan ventilation will be provided for all enclosures containing gas heaters. No plastic hose will be allowed for gas supply, and tanks will be kept outside enclosures.

5. Housekeeping

Tripping hazards will be eliminated by removal of hose, cables, and ropes from walkways, by proper storage of materials, and by disposal of waste material.

All debris shall be removed daily.

Nails shall be removed from lumber or flattened immediately.

Oil and grease spills will be cleaned up immediately, and icy, slippery surfaces will be cleared and sanded.

Horseplay

No horseplay will be allowed. Actions of employees will be controlled by foremen or superintendents.

Horseplay will be cause for discharge of offending employees.

6. Individual Protective Equipment

Workers will be thoroughly indoctrinated in the instinctive use of required protective equipment.

All personnel will wear hard hats throughout the project.

Goggles will be worn for all chipping, grinding, drilling, use of acids, chemicals, epoxy, power actuated or impact tools, and concrete and masonry saws.

Ear protection will be worn by operators, drivers, or mechanics exposed to continuous high-level sound intensity.

Respirators will be worn by all persons exposed to dust or chemical fume inhalation, spray painting, or sandblasting. Assistants as well as operators will be protected.

Safety belts, lifelines, and lanyards will be worn by all workers on elevated locations not protected with railings on scaffolds and platforms, including boatswain's chairs.

Impermeable rubber or plastic covering will be worn to prevent skin contamination by bacterial infections, acids or chemicals.

Non-slip soles will be worn on safety shoes.

Fluorescent vests or belts will be worn by employees exposed to traffic by night or day.

7. Ventilation

Positive means of ventilation including exhaust fans and ducts, shall be provided during work with enclosed gas heaters, solvents, resins, and other toxic materials in confined spaces to reduce concentrations below toxic limits and to remove flammable vapors.

8. Hand and Portable Tools

All tools shall be in good condition without mushroomed heads and split handles, repaired promptly or removed from the site. This includes privately owned equipment of the workmen.

Tools shall not be left overhead to fall. Throwing of tools is prohibited.

Blocking or cribbing shall follow closely all lifts with jacks.

Guards shall be installed over all cutting, rotating, or moving parts.

Safety lashing shall be installed at pneumatic hose connections.

All powder-actuated tools, all charges and studs will be stored under lock and key. Use of such tools shall be limited to operators possessing certificate authorized by manufacturer's representative after training in safe use and maintenance of that model equipment. Drive no stud within four (4) inches of a masonry corner.

9. Temporary Construction

Ladders shall be sound and solid, long enough to project three feet over top landing, and be secured top and bottom against falling.

Trailer entrances shall have steps, not blocks, and handrails.

No single plank shall be used for staging or walkways and will not be supported on ladder rungs.

All cribbing and blocking will be secured in position.

All scaffolding shall be erected of sound materials, securely braced, and provided with guardrails and toe boards to prevent falling of material or dislodging tools.

Every opening in a deck or floor will be closed with solid covers or surrounded with a rigid barricade.

Excavations, trenches, and openings shall be barricaded and marked with lights during hours of darkness.

Adequate lighting will be installed for stairs, ladders, unlighted compartments and walkways during work periods. Minimum intensity shall be five (5) foot candles.

General Job Conditions

Runways, ramps, platforms, and scaffolds over four (4) feet in height shall be guarded with handrails and toe boards.

All oxygen and acetylene cylinders shall be handled with care, protected against excessive heat, including the continuous direct rays of the sun, and accumulations of ice and snow.

a. Cylinders containing oxygen shall not be stored alongside combustible materials or cylinders containing combustible gases.

b. The protection caps shall be in place when cylinders are not in use.

c. All compressed gas cylinders in service shall be secured in substantial fixed or portable racks or hand trucks. Upright cylinders will be adequately secured against displacement.

The practice of throwing tools from one location to another or from one employee to another or dropping them to lower levels shall not be permitted. When it is necessary to pass tools or materials under the above conditions, suitable containers and/or ropes shall be used.

No work will be done from top step of a stepladder.

Structures will be braced against forces in all directions until able to stand alone, and will not be overloaded with material stock piles.

10. Electrical Hazards

All installation, temporary as well as permanent work, shall comply with the National Electric Code and shall be installed by licensed electricians.

Ground-fault circuit interrupters will be used or an assured equipment grounding conductor program will be established on all 120 volt, single phase, 15 and 20 ampere receptacles that are not part of the permanent wiring.

Portable electric generators shall be properly grounded to ground rods or water lines. Electric welders shall be bonded *and* grounded.

Portable electric tools shall be grounded with three-wire cords and receptacles, or by bonding wires to low-resistance ground and leads unless of Underwriter Laboratories approved double-insulated type.

Temporary electric wiring shall be suspended overhead.

Lamp bulbs shall be guarded.

Bypassing of protective devices will not be tolerated, and switches and boxes shall be closed. No work will be performed on "hot" lines of any voltage.

11. Safe Clearance Procedure

Before repair or adjustment of any mechanical, electrical, pressure, or hydraulic systems, inadvertent operation will be prevented by locking of switches, controls, or valves or moving parts. Authorization for interruption of systems will be obtained by advance preparation of approved schedule of work, identification of responsible supervisor of the work, and submission of safe operating procedures. Authorization for safe clearance will be obtained in advance from the supervisor, and no system will be interrupted without approved safe clearance procedure.

Ramps, Runways, and Platforms

a. Ramps shall be of sufficient width for the equipment using them.
b. Walkways or platforms used for erection of equipment above the ground will be equipped with guardrails.
c. Walkways along the boom of the large draglines shall be of sufficient width with cleats on 15 inch centers and an approved guardrail along the outside edge with the boom forming the inside rail.

Excavation and Trenches

a. Excavations made by the use of power shovels or backhoes, loading to off highway units and large draglines casting directly to spoil banks. All equipment will be equipped with modern safety devices and will be constantly checked to avoid mechanical failures.

12. Powered Equipment

All machines shall be inspected prior to work on the project by the operator or mechanic.

All machines will be examined daily for safety appliances, and conditions and all defects repaired promptly. Periodic maintenance schedules will be followed.

Crane and hoists will be given static load test of 125% of the maximum service load. In the cab of each crane, signs for maximum loads at all boom angles and warning of ten foot clearance at overhead wires will be posted in operator's view.

All compressors will be given a hydrostatic test of 125% maximum working pressure biannually, and the certificate posted on the project.

Rigging, hooks, slings, and tackle will be examined frequently and defective items removed from the site. All hooks will have safety latches. Worn cable and old socket fittings will be replaced.

Not less than three (3) clips will be used on cable returns applied correctly.

All machines will be shut down for lubricating or oiling. During repair of all machines, blocks or stops will be set to prevent falling or moving of parts should any hydraulic line or control device fail.

Workers shall climb carefully with handholds and grab irons, not by jumping on or off any machine, and in no case while the machine is in motion.

All signals to operators shall be given by designated, trained signalers.

Rotating parts, gears, cutters, and coupling will be covered by guards.

The contractor shall comply with the manufacturer's specifications and limitations applicable to the operation of all cranes.

All tractors, graders, forklift trucks, dump trucks, truck cranes, and the like will be equipped with approved reverse signal alarms sounding automatically.

Tractors, dozers, front end loaders, and graders will have effective rollover bars and seat belts, with plate mounted certifying compliance with SAE standards.

Heavy hauling units and trucks over five tons will be equipped with emergency brakes automatically stopping the machine if service brakes should fail.

Dual wheels on vehicles will have stone ejectors between each pair of tires.

Track and wheel vehicles will be cleaned of all earth by water under pressure, and all hand tools will be cleaned of earth by wire brushing prior to removal from the site.

No one will be permitted to ride in any truck body, crane hook, or bucket.

All truck bodies carrying loose materials will be covered.

Fuel truck will be bonded with a cable to machine being fueled to prevent static discharge.

All repairs of hydraulic systems will be with new manufacturers' parts.

Mortar mixer hopper will be screened with grille having maximum two inch spacing.

Fork lift or other internal combustion powered machines will be used indoors only when equipped with certified exhaust devices.

Circular saws will be guarded, and swing saws will have retracting spring action guards.

Solid and level standing will be provided for truck cranes and readi-mix trucks.

Roofer's hoists will be erected with anchorage secured to structure, not cantilevered against weight of materials.

Elevating conveyors will be secured in position, men will not ride on belts, and two inch thick roof will be installed below belt at building entrances.

Crane booms and hoist lines will be kept at least ten feet away from all overhead wires.

Quick shutoff controls will be installed on conveyor belts.

13. Steelwork

No person shall work on steel members when there is a film of moisture or ice on surface of the steel.

Tag line will be used to control swing of crane lifts.

No part of structure will be loaded with weight that will endanger its safety or stability.

Use temporary bracing and guying to support all steel framing until structure can stand alone.

Concrete dowels will have ends protected by plastic caps, as will electrical conduit stubs.

Workers will not use roof joists and beams for temporary support without platforms, staging, scaffolds, or walkways.

14. Safety Nets

All work not done from scaffolds installed as described above or from boatswains' chairs will be done only after protective safety nets have been suspended below. Net shall extend eight feet beyond the edge of the work surface where employees are exposed. Maximum drop will be 25 feet.

All new nets shall meet accepted performance standards of 17,500 foot-pounds minimum impact resistance as determined and certified by the manufacturers, and shall bear a label of proof test. Edge ropes shall provide a minimum breaking strength of 5,400 pounds.

Repairs will be made promptly to any part of the system damaged.

Maximum size of net mesh shall be six inches by six inches.

Cable connections to metal through holes, rings, other cable or anchorage will be made with shackles and all loops of cable will be protected with thimbles. Not less than three cable clips will be used on any cable connection, applied correctly.

Condition of nets and supporting system will be inspected daily by a licensed rigger.

No work will be done from net and no workmen will stand or sit in the net for any purpose except its inspection, cleaning or collecting debris, or repair.

Perimeter cables will be tensioned to reduce sag in the nets.

Net will attach to cables only with safety hooks, and cable will be steel of adequate strength.

15. Suspended Scaffolds

Installation, relocation, or removal of suspended scaffolds will be approved in advance. Drawings of suspended scaffold and all associated equipment will be approved in advance of use.

Enclosed scaffold machines will be inspected in advance by a licensed rigger, dismantled, lubricated and reassembled with a tag bearing the date and name of rigger while in use on the project.

Movable scaffolds will be load tested to 150% of maximum service load.

No ladder shall stand upon a suspended scaffold.

Two or more suspended scaffolds shall not be combined by bridging between planks or ladders, whether secured to scaffold or not.

Outrigger supporting beams will be steel, bolted to structural steel members and not to punch-outs in concrete deck. Cables will be prevented from slipping off the ends of outriggers by fixed stops.

All movable scaffolds will be sway braced and enclosed with standard guard rails and toe boards.

For attachment of safety-belt lanyards, a lifeline reaching to the ground will be independently hung for each person on the suspended scaffold, float, or boatswains chair.

16. Asbestos and Hazardous Waste

All work with asbestos and hazardous waste will be done only by workers certified to work with these materials.

17. Traffic and Project Vehicles

Drivers will be continuously supervised throughout the project to respect the rights of the public and to drive defensively in view of the serious hazard. A positive stop will be required of all project vehicles at all street intersections and at project entrances and exits.

Standard traffic warning signs, barricades, markers, and lights will be set for protection of workers and the public. All traffic controls will be maintained with daily inspection and repairs, including all lighting.

Blowing of dust from the project limits will be prevented by wet drilling, control of exhausts, and approved use of dust palliatives.

Uniformed police officers may be required at critical locations and during rush hours. Police will be coordinated with city officials to prevent accidents.

Barriers will be set to prevent vehicles overrunning the edge of excavations or embankments.

18. Epoxy Handling

Furnace shields will be worn by workmen during mixing and placing.

Protective coveralls and neoprene gloves shall be worn by all those handling epoxy materials.

Personal cleanliness will be required at all times, with showers, wash basins, soap or skin cleaners, clean towels, and change of coveralls provided at all times epoxy materials are in use.

Protective ointment will be used as recommended by epoxy material.

Fire extinguisher will be used at the location of all mixing and placing.

Deck areas subject to epoxy contamination will be protected with paper which will be disposed of when stained with binder or mix.

Smoking is not allowed within 50 feet of mixing and placing, and no vehicle or machine powered by gas engine is permitted to stand within 50 feet.

Contaminated clothing, wipers, cleaning cloths, and papers will be disposed of each day in an approved manner.

Clean rags or paper towels will be furnished for absorbing spills or wiping contamination from epoxy. Cloths and papers will be disposed of in an approved manner each day.

Workers using epoxy materials will be instructed in safe use of the systems, including hazard from contamination, necessity for cleanliness, and disposal of waste materials. Each person will be instructed in first aid measures against contamination, especially the flushing of the eyes.

Portable eye wash will accompany each use of epoxy; use of a hose is not permitted.

19. Demolition

Bracing in all directions will be installed to resist forces and provide support for adjoining work during demolition.

Permanent work will be protected from damage by enclosures, coverings, and guards during demolition.

Debris from demolition will be removed from the site immediately.

Utilities will be cut off, marked with tags, and locations recorded before undertaking demolitions.

20. Safety Signs

Safety signs will be erected by the company at the office to assist in training and encouragement of accident prevention consciousness.

21. Emergencies

In the event of warning of severe storm, personnel will be evacuated and equipment secured as directed by the superintendent to prevent losses or damage.

In the prospect of a thunder storms, all work in open areas will be suspended.

During tornado watches, equipment will be removed from exposed positions and personnel will be protected. Crane booms will be laid down.

22. Accident Reporting, Analysis, and Prevention

Every accident will be reported to the respective supervisor.

Supervisors at all levels will follow up with training and observation directed to the prevention of repetition. Weekly safety meetings will review accidents and discuss remedial action by all workers.

23. Occupational Safety and Health Act of 1970 (OSHA)

The Volpe Construction Co., Inc. developed this comprehensive safety plan to comply with the OSHA requirements. Up dates are required from time to time as the Act is amended or as conditions may change.

24. Coordinator of Program

Director of safety or company employee appointed by the President.

Setting up a safety plan is a task that should not be compromised. Begin by appointing a safety engineer or safety supervisor. In smaller firms, someone working in another area may be assigned the responsibility. Avoid selecting the project manager, superintendent, or foreman for this job, because they are on the job every day and are too close to see all the hazards.

In a larger firm, the safety supervisor helps the superintendent and foremen organize a plan for safety and accident prevention at the start-up of each job.

A key to any good plan is continuous vigilance. Bad habits creep back in quickly if the safety supervisor is not visiting the jobs on a regular basis.

Our company defines our safety program by what it hopes to accomplish and why we think safety is important.

11.2.2 Safety Awards

One effective plan provides awards to both superintendents and foremen on a quarterly basis where there have been no accidents resulting in lost time for any of their workers. Supplementary prizes are awarded at six months. An annual prize of a larger amount is something all strive to achieve. Eligible foremen include those for each trade employed by the firm—carpenters, laborers, bricklayers, cement finishers, etc. Where records are equally good, the foreman who has supervised the greatest number of man-hours gets the award. If the project engineers participate in the plan, they too receive awards.

Awards are presented at dinner meetings. There is always some good-natured ribbing by the winners. This often helps those who did not get an award realize they were not paying proper attention to safety. By working for better results, they too can participate in awards at the next meeting.

When top management turns out for the meeting and the "boss" makes the awards, the employees know he is interested. They know he is aware of which superintendents and foremen have the best safety performance. Other plans can award prizes other than cash awards.

11.2.3 Job Safety Inspection Report Form

On a large project the safety engineer will tour the job each day and report findings to the superintendent. A good checklist report form is shown in Fig. 11.2.3. Note the form heading includes the number of accidents to date. The costs of the accidents to date are also listed. This is a figure from a formula that you establish. Items to be inspected are listed with space for comments. The superintendent signs the form to acknowledge the conditions found.

The report is routed through the project manager, general superintendent, operations vice president or president, and back to the safety engineer. He can review any notes or comments by higher management. Note at the bottom of the form there are places for those others to indicate they have seen it.

Lost-time accidents to date and lost-time days to date are shown separately. Facts about dangerous conditions can be listed under Remarks. Improvements or worsening of conditions since the previous inspection can be noted.

Each foreman ought to have a weekly five-minute toolbox meeting with his crew, either at starting time or right after lunch. He discusses the safety performance of the previous week and reviews safety methods that will be used on upcoming work. These safety review meetings are the backbone of the safety program because they directly involve the people who are really responsible for the safe job—the workers.

This form can also be used as a safety checklist by a smaller organization where the safety program can be informal, but effective.

When management reviews these forms and finds continual unsafe conditions, action should be immediately taken. Talk to the superintendent and foremen if necessary.

THE VOLPE CONSTRUCTION CO., INC.   SAFETY REPORTS

PROJECT_____ DATE_____

NO. OF ACCIDENTS _____   COST OF ACCIDENTS_____

CYLINDERS   HOUSEKEEPING   POWER TOOLS

_____   _____   _____
_____   _____   _____
_____   _____   _____
_____   _____   _____

ELECTRICAL   HEATERS   PROTECTIVE EQUIPMENT

_____   _____   _____
_____   _____   _____
_____   _____   _____
_____   _____   _____

EXCAVATION   LADDERS   RAMPS

_____   _____   _____
_____   _____   _____
_____   _____   _____
_____   _____   _____

GUARD RAILS   MOBILE EQUIPMENT   OTHERS

_____   _____   _____
_____   _____   _____
_____   _____   _____
_____   _____   _____

FIRE PROTECTION   MEDICAL   STAGING

_____   _____   _____
_____   _____   _____
_____   _____   _____
_____   _____   _____

REMARKS_____
_____
_____
_____
_____

SUPERINTENDENT'S SIGNATURE_____

ACKNOWLEDGEMENT: PM_____   H.G_____   JEB_____   RWN_____

FIGURE 11.2.3 Job Safety Inspection Form

11.2.4 Pictures of Hazards

Unsafe conditions can be shown in candid photos taken by your safety supervisor. These pictures are put in a booklet with a short story. When this is shown to the superintendent and foreman on the next visit, it is amazing how real the hazard seems.

Slides of the various unsafe conditions can be projected at a safety meeting. Your superintendents and foremen analyze them and recommend corrective actions. This dramatizes unsafe conditions.

11.2.5 Interest from Others

Your insurance carrier is a good program source for safety meetings. His safety person can get movies that suit your needs. Owners and architect representatives are usually interested in what you are doing and may be able to add to your program. Invite them to your safety meetings. This will also make them aware of your efforts in the safety field. They will consider your firm well organized and capable of providing good value on either contract or negotiated work. Your insurance carrier's safety engineer should occasionally make job site visits with your safety supervisor.

11.2.6 Cost of Safety

An organized safety program may sound expensive but you can tailor the program to the needs of your firm. It will pay for itself in both dollars and satisfaction. Safety is not expensive when it keeps you in a better competitive position.

11.2.7 Occupational and Safety and Health Act

This law affected all direct federal work and all federally assisted work programs. It is a broad act administered by the Department of Labor. It has been expanded and today covers most work. You must keep up to date on this and the state laws affecting safety. Penalties for noncompliance can be rather expensive as well as time consuming.

It is important that you be aware of its provisions. Make sure that you have a copy of the law and its interpretations available.

11.3 VARIOUS INSURANCES

A contractor must provide many types of insurance for his own and his client's protection. You need to know each type and how it relates to the work your firm performs.

11.3.1 Employer's Liability

Safety on the job is ordinarily related to worker's compensation. In some states it is possible for an injured worker to sue for his injuries instead of going through the normal channels of worker's compensation coverage. This holds true for a subcontractor's employee under some conditions. Employer's liability, or Coverage B as it is sometimes known, protects your firm for these claims.

11.3.2 Property Damage

Damage done to property of others is generally covered by a property damage policy. Walls or footings of an adjoining building may be broken or otherwise damaged by your forces in spite of careful planning.

Beware of the exclusion clause when a building or property is "under your care, custody, or control." If the owner vacates a building so that alterations can be made, the contractor has care, custody, and control of the property. Damage is not covered by the usual insurance and cost could be considerable if a crane drops a beam through the roof. Consult your insurance carrier for the needed coverage.

11.3.3 Public Liability

As the term public liability implies, this type of insurance covers bodily injury to the public resulting from the construction operation.

11.3.4 Independent Contractors

Damage caused by an uninsured or worker's subcontractor or his employees is covered by a separate policy or clause relating to independent contractors working for you.

11.3.5 "Contractual Liability and Hold Harmless" Coverage

Contractual liability insurance has become very technical and should be discussed with a technically qualified agent prior to bidding on the project or before signing the contract. A lot of problems can arise from the so-called "hold harmless" clauses that are sometimes written into contract specifications. Under this coverage you may be assuming liability for the owner or a third party. Some specifications make this coverage so broad that it is almost impossible to figure out exactly what you are covering. Some architects are attempting to extend this to include errors and omissions of design. You may end up involved in some action in which you had no concern at all.

It is imperative that you study this clause and check it out with your insurance carrier. He will advise if coverage is available and what it will cost. Make sure you do not sign a contract with a "hold harmless" clause you cannot insure.

11.3.6 Completed Operations

Insurance covering a completed operation is sometimes specified. Many firms carry it on all projects regardless of specifications. This protects you in the event of injury to the public or damage to property because of negligence in a completed building. For example, a building is completed and has been in operation for a year or so. One day a portion of the ceiling falls on people and merchandise causing injury and damage. A lawsuit could develop. Without completed operations coverage you would have to defend yourself and pay any judgments.

11.3.7 Automotive

Make sure your automobile policies cover cars and trucks both on and off the highway. Some policies only cover operations on public roads and streets. Territorial coverage is needed for off-highway and private property operations. Be sure the limits are adequate for all coverages.

11.3.8 "All Risk" Builders Risk

"All risk" builder's risk or regular fire and extended coverage are other important policies. Whenever it is specified that the owner will carry this policy, make sure of the coverage he will provide and protect yourself for the rest. Some owners when purchasing builder's risk insurance take on a sizable deductible of $1000 to $5000. If the owner does not protect you for this deductible, you have to cover that amount. This can be done by taking out a policy of your own or by carrying sums in your estimate to cover this loss. You should talk to the owner to make sure he understands you will be adding sums to your bid to protect yourself and if he were to reduce the deductible, it would be cheaper for him in the long run. If no mention is made in a specification about this type of insurance, it is your responsibility to protect yourself against the risk. Do not feel safe just because there is not a specific requirement for you to provide any form of "builders risk" insurance. Such an obligation could well be hidden in the "responsibility of contractor" clause. "All risk" builder's risk generally covers most of the perils involved, whereas fire insurance and extended coverage are more restrictive. Check your specification for the type to be provided.

In addition to fire, "all risk" builder's risk can cover damage from vandalism, wind, riot, aircraft, flood, and explosion. (Watch out for boiler coverage—it may have to be separate.) This coverage may be purchased as a completed value in which the premium is based on the total insurable value of the project at an average rate. The other method is the reporting type wherein, as the work progresses, you increase the coverage.

The reporting type is increased in the amount of each requisition as each is approved. If someone neglects to increase the policy, you are underinsured. If a loss occurs during this period, any costs in excess of the insurance coverage are borne by you. The completed value gets the best recommendation, as it leaves nothing to chance, which might occur under the reporting type.

11.4 HOW TO EVALUATE YOUR NEEDS

A question that arises time and again concerns the limits that are desirable for each type of coverage. Specifications sometimes set the limits. However, these limits can be too low or too high. Do not worry about the high limits, but beware of the low ones. Firms have been put out of business as a result of a court suit in which coverage was inadequate.

Important factors to consider when setting limits are the proximity of adjacent buildings, their value, condition, etc. Other factors include nearness of public ways, traffic volume, foot traffic, and other exposure requiring insurance protection. Limits should be high enough to protect all possible losses.

11.4.1 Umbrella Liability

An umbrella liability policy provides protection beyond the basic limits of your comprehensive general liability and automobile liability policy coverage. The underlying limits you must carry are not a standard amount because they vary with each carrier. There is a trend toward higher limits being required in your basic policy by the umbrella carriers. These types of policies are required by some owners in varying amounts. Another factor to consider in buying this type of coverage is the high cost of legal defense which these policies generally cover.

11.4.2 Group, Life, Key Man, Pension, and Profit Sharing

Group life insurance is available to you and your key employees up to certain limits, and premiums are a tax-deductible expense. This coverage is on term, paid up, or a straight life basis. Individual life insurance policies can be carried in order to purchase shares of stock from estates, etc.

Insurance for key personnel protects the company when an important executive passes away unexpectedly. This policy provides funds to offset financial losses that may result until a suitable replacement is found.

Pension and profit sharing plans can be arranged through any of the insurance companies as well as many different types of financial managers.

11.5 EXPERIENCED AGENT AND A QUALIFIED CARRIER ARE ESSENTIAL

The importance of a good, reliable, experienced agent cannot be overstressed. He in turn must represent a strong and responsive insurance company who can help determine the best and most economical coverage for each of your projects. You simply cannot go along with a broker friend who is unfamiliar with coverages as they apply to the contracting business. You may find yourself in trouble without proper coverages as a result.

Insurance policies are constantly changing to meet the changing needs of contractors. Be sure your agent keeps you up to date so that you can take advantage of the latest economies and protections offered by progressive insurance carriers.

11.6 INSURANCE CHECKLIST

The following is a good checklist to use regarding coverages needed. The specifications give the actual requirements for each project. Your insurance agent should also review the specifications to make sure you have the proper coverages.

1. Worker's compensation (according to statute)
   A. In all states where operations are conducted
   B. Employer's liability (Coverage B)
2. Liability
   A. Automobile
   B. Comprehensive, covering premises and operations
      a. Independent contractors
      b. Contractual
      c. Completed operations
      d. Coverage for X (blasting), C (collapse), or U (damage to underground facilities, if necessary)
      e. Personal injury coverages, such as false arrest, invasion of privacy, libel, assault
3. Property insurance
   A. Contractor's equipment, including rented or leased
   B. Automobile physical damage
   C. Fire insurance on owned, rented, or leased property or property in your care, custody, or control
   D. Builder's risk, especially on an All Risk Basis
   E. Boiler coverage, if necessary (this is generally carried by the owner)
4. Miscellaneous
   A. Fidelity, forgery, credit card
   B. Holdup and associated money coverages
5. Group, life, key personnel, pension, profit sharing

11.7 CONCLUSION

Safety is an important aspect of your business. It improves the working conditions of your employees, which results in better production and a safer work environment. These factors make better workers want to work for your firm.

Insurance is a subject that is fairly complex with all the coverages available including the hold harmless clauses required. Insurance policies change and vary from state to state. You should spend time with a good insurance agent and have him explain which policies cover what and the coverages available. A good agent who is experienced in construction coverages is a must. Coverages for most projects are generally standard but on occasions a different twist is added. Your agent should review these requirements so you will know if coverage is available and thus can adequately protect your company.

CHAPTER 12

Marketing

12.1 INTRODUCTION

Marketing is a very important part of any business whether it be construction or selling widgets.

In construction, marketing is selling your company services to potential customers. Before someone starts to market, he must prepare to do so. Unless he develops a business plan, he cannot prepare an effective marketing plan.

A good brochure showing the previous work done by the firm and the individuals employed in the firm is a must.

If you will review the proposals that are submitted, particularly those to private owners, you can see what has been requested for submissions.

These requests are the things an owner looks for in a construction company. Reviewing one of your proposals does not reveal all the items you should emphasize, but when looking at all of the proposals you have submitted, you will notice a pretty good trend in the information required.

Therefore, it is important that these facts be put forth in your selling effort. Another important item is recommendations. When you complete a successful project, do not be afraid to ask for a letter of recommendation to put in your selling kit.

Owners generally know each other in a given area and they feel comfortable calling one another to check out a contractor's capability and reliability.

12.2 CALLING ON ARCHITECTS/ENGINEERS

In addition to calling on owners, it is also important to call on architects/engineers who many times have a great deal to say about who will be the selected contractor or who will be on the list of bidders. It is therefore, important to call on them when seeking work to get them interested in your firm. Sometimes they may not have anything ready at the moment, but if they are convinced you represent a good firm they will remember you for consideration in the future.

Most construction firms have a strong suit such as on-time delivery of the project which is important to an owner, saving interest money on the construction loan or allowing them earlier use of the project. If it is a shopping center or other building where people are to be employed, this on-time delivery means the people they have hired will become productive at once and not be a burden to the owner who, if he has hired people in anticipation of the expected completion date, must pay them while waiting for the project to be completed.

12.3 PUBLICITY AND COMMUNITY RELATIONS

We mentioned that being a part of the community is important. This, too, is a selling tool. As mentioned, people get to know you in a relaxed atmosphere which allows them to make the right judgment about you as an individual. This comfort with you will carry over into respect for the firm.

It is good to review publicity developed by other firms in your area. This will give an idea as to what they are promoting. It also will give an indication as to what makes good construction publicity for the firm.

If the firm gets an award or has a safety meeting, an employee receives a promotion, or any other newsworthy event occurs, send out a press release to the local newspapers. This, too, is a way to get the firm name out to the public. It makes people aware of the company, which can make marketing easier.

12.4 PUBLIC RELATIONS FIRMS

Hiring a public relations firm can be a tremendous help. This is what they are in business to do—promote firms to improve their image. They help develop brochures that tell the story you are trying to get across. They help prepare press releases for publicity purposes. They generally know people in the media, which helps get the news stories they generate published. These stories can be about the award of a new contract or the opening of a building the firm just completed. It can also be about a new system you are employing or an unusual feature about a project you are constructing. These are some general ideas on publicity stories that can be generated.

Whatever you do, make sure it is done in a professional manner. An effort that is halfhearted will give the wrong impression about the firm.

12.5 ASSOCIATION MEETINGS

Attending association meetings will also give ideas on what makes for good publicity. Most associations have a marketing committee, that helps develop ideas for marketing. They develop the type of brochures that make good sense for marketing.

12.6 MANAGEMENT SUPPORT

A very important factor is that management must support the marketing effort in a very strong way. All employees must help sell the good qualities of the firm, which is a subtle marketing point. Their actions reflect on the firm.

12.7 SUBMITTING PROPOSALS

When submitting proposals for projects, they should include your brochure and the pertinent information referred to in the Request for Proposals (RFP). Generally these proposals include information on the key people who will be involved on the project being considered. Make sure you have up-to-date resumes for these people. If the request is for a particular type of project, list the people who have experience in this field. Do not be afraid to include the letters of recommendation that were referred to earlier. Do not try to hedge in giving information for this will be very readily noticed.

12.8 REFERENCES

A good book in the Wiley series that refers to marketing is *Construction and Engineering Marketing For Major Project Services*, by B. C. Gerwick, Jr.

12.9 CONCLUSION

Marketing is a very important function, and one to which a lot of attention must be given. You have to get out and sell your firm's capabilities to owners and architects. Some people are naturals at selling whereas others find it hard to do. It is important to project a good strong image. Having a good track record and reputation makes marketing a much easier task. If owners know you or of you through a good reputation, this helps tremendously. This is another reason for being active in community and industry affairs. You meet people who can judge you better and are able to judge your character. They may also know those you do want to business with for reference purposes. If you are not known or do not gain a good reputation, it is much more difficult to open the doors seeking new work.

CHAPTER 13

Trade Associations

13.1 PURPOSE

Membership in trade associations offers many benefits to an individual firm. You learn about common problems and the best solutions from other members of the industry.

Associations keep watch over the broad spectrum of industry both locally and nationally on such items as legislation, labor, contracts, subcontracts, and relationships between owner, architect, engineer, and contractor.

Laws, both good and bad, affect our industry. A unified stand is most effective in preventing passage of injurious legislation or helping the passage of good laws. The national office of an association aids this process by alerting local chapters so that members can inform congressmen and senators about the impact of pending legislation on the construction business.

13.1.1 Types of Associations

For the general contractor, there is at least one local chapter of the Associated General Contractors of America in every state. These chapters are all affiliated with the national association and therefore are a very strong group when it comes to getting some point across to either the government or other organizations. For specialty contractors, there are various trade groups in electrical, mechanical, heating and ventilating, concrete, asphalt, steel, etc.

Local groups not affiliated with national organizations are strong and effective in many areas. However, they sometimes miss the big national picture. They do not have the strength in congress needed to fight bad federal legislation when it is introduced. Conversely, they have a problem getting legislation through that is needed.

In addition to trade associations, there are professional organizations such as the American Society of Civil Engineers and the American Institute of Architects and others that actively follow legislation and other problems that can affect your business directly or indirectly. Where you or members of your staff are trained in a professional field, it is advantageous to maintain your membership and contacts with these professional groups.

13.1.2 Participation

Participation is essential to get your point of view across to others. It is easy to sit back and criticize. But when you avoid participation, all the facts may not be known, and the wrong impression can be promoted by your association. By participating you help sponsor the things you feel are good for the industry. If you get elected to the board of directors or as an officer, it is amazing what you learn about the workings of the association and the important work they do. You get an appreciation of the work needed to accomplish something of importance. When giving a speech as national president of A.G.C.A., I made the comment that "association work is like a bank account. The more you put in, the more you get out of it." This is still true today.

13.2 LOCAL CHAPTER MEETINGS

Meetings are planned for the general interest of members, and attendance will usually provide new ideas of techniques that save money and keep you competitive. You never gain the same knowledge by simply reading the speaker's text at a later time. The best idea for you may be gleaned from looking at the speaker's slides or listening to his answer to a pointed question.

Meetings of business groups invariably involve competitors and are thus useful in keeping up with changing trends.

13.3 LOCAL COMMITTEE WORK

Committee work can be most time consuming, but also most rewarding. You find that committees are usually made up of active, busy, and knowledgeable people like yourself. You get to know firsthand the people who are creating the changes in the local construction industry.

13.3.1 Solving Problems

Committee action is impersonal so that no one is singled out for acting out of selfish interest. For example, a local committee can meet with a public body whose contracts are demanding, unreasonable, and without provision for arbitration. An open discussion of each side of the problem can develop better understanding and may lead to a satisfactory solution.

You get to know each other and gain mutual respect. This helps in any further negotiations and would be difficult to accomplish as an individual firm.

Unrealistic clauses in specifications by architects and engineers meets with their counterpart.

13.3.2 Unified Action

A clarification or request for certain changes in bidding documents is more easily obtained when an association chief executive contacts the owner, architect, or engineer. An association request generally carries more weight than an individual request.

Your position for labor negotiations will be stronger when all contractors get together and establish good basic contracts with reasonable wages and work rules. You may disagree with certain actions, but the majority of group action will be to your advantage.

13.4 NATIONAL ORGANIZATIONS

At the national level, ideas, problems, and requests are handled by an experienced staff. Useful information is gathered from local chapters and sent out to all members. A national staff gets to know the heads of labor unions, federal agencies, and professional societies, and this contributes to a more smoothly flowing industry in which problems are handled more easily. For this reason information often flows from a local chapter to the national organization, which in turn informs the other locals of the problem and its possible solution. One thing a local chapter must remember, however, is that the national organization is looking at the overall picture; if forty-nine states require an action, the fiftieth state cannot expect to stop its progress for its own selfish purposes.

13.5 NATIONAL COMMITTEES

National committees are generally counterparts of your local committees. They usually meet during the annual convention and several other times during the year. They deal with problems on government contract awards, A.I.A. forms, specification and payment clauses of government agencies, and other national problems.

Some of the committees are established with other national associations as well as federal agencies to develop many of the contract forms used in our industry. The reason these joint committees are formed is to try to develop standard contracts forms which fit different types of construction. It is simple to write a contract without regard for the other participant and what affect it will have. It is far more important to write a contract that is fair to all parties. A contract must be one that both parties can live up to and not so one sided that it becomes impossible to accomplish. If the contract is too one sided, it cannot be accomplished or is so onerous the courts could very well throw it out on these grounds. This is where your involvement on committee work is important. You are not only able to represent the contractors but help the entire industry come up with good contract forms. Here is where good communication is important to both sides by having an opportunity to understand what the words mean from both sides, thereby helping make a sound judgment. It is easy to sit back and criticize what has been done if you do not participate. Lack of input even by an organization can cause a lot of problems.

These meetings give you a further benefit by being able to sit with architects and engineers with who you may not otherwise have had an opportunity to see in your everyday routine. This too is important to your future.

Committee meetings are usually open to all association members; they can observe and speak, but cannot vote on committee actions. Participation on a national committee provides benefits to you and your firm. One valuable idea gleaned from attending the annual convention can easily make the cost of the trip worthwhile.

13.6 CONCLUSION

Many benefits are derived from belonging to the local chapter of a national association. You save time and money in solving problems through the unified action of an association. Work is spread out among the many skilled association members so there is little duplication of effort.

To me, an important benefit of association work is learning about problems before they occur. Coping with these problems is easier when you have been forewarned and are familiar with successful solutions.

When you do not participate locally and nationally in shaping the destiny of our great industry, both you and the industry are losers, with the heavier loser being you.

C H A P T E R 14

Joint Ventures

14.1 REASONS FOR JOINT VENTURES

Joint ventures are fairly common in the construction industry. Joint ventures orig-inated on large projects for two reasons. First was the need to assemble the required bonding capacity and second was the objective of spreading the risk of a large project, much as insurance companies do when they reinsure parts of a large policy.

When your bonding capacity is spread thin on current projects, and a good profitable project of several million dollars might be obtained by negotiation, getting a partner for a joint venture is often the answer.

14.2 HOW THEY WORK

Joint ventures are formed by two or more firms. One firm is designated as the sponsor. This firm is responsible for the operation of the project. As a rule the sponsor has the larger percentage of the venture. The project is controlled by a committee of the venture partners in much the same way that a board of directors controls a corpora-tion. Committee members include the principal of each firm or his designee.

Generally, each firm participating will take off quantities and estimate the cost of the work they will do with their own forces. These figures are compared and if they vary substantially, the difference in quantities must be reconciled first. Then cost estimates are reconciled, which confirms the accuracy of the estimate.

Available personnel are listed by each firm. From these lists the job organization is set up. In some instances, the sponsor may have sufficient people to operate the project without help from the other partners. There also may be a case where a special type of work is required and one of the other partners may have that capability. This is another reason why joint ventures are formed.

Equipment needs are determined. Each partner provides a list of available equipment. The equipment used is charged to the joint venture on a rental basis if it is rented from one of the partners the rental rate would be the same as if rented or leased from an outside source.

Working capital to finance operations is advanced by the venture partners. When the amount of operating funds is decided upon the money is put into the joint venture bank account on the basis of each partners percentage of participation.

Profits or losses are shared on this same percentage basis. However, the liability can be greater than this percentage if one or more of the other partners suffer serious losses and go bankrupt. All partners are generally responsible on a joint and several basis, which will add to your exposure if one of the partners fails.

The joint venture committee meets periodically to establish policy, review progress, and examine costs. When there is disagreement, the sponsor has the final say and can proceed as he deems best. Generally speaking this does not occur because you should not go into a joint venture unless you know the other partner on a good solid basis. If in fact problems do occur the other partners have recourse through arbitration or other means that are spelled out in the agreement if they so desire.

14.3 SELECTION OF A JOINT VENTURER

Selection of the joint venturer is the key to the success of this type of work. Partners must know, respect, and trust each other one hundred percent. Any doubts as to the ability or integrity of a proposed partner should be brought out into the open at the outset. If you have serious doubts, do not participate. Once an agreement is made, it is too late to back out.

Each partner should be selected on the basis of what the firm can contribute to the project. The venture should be assembled to have strength in all the needed areas of the project.

Bonding capability is more easily established when the proposed partner is known to your bonding company. The financial position of each firm must be made known to the other. Where a proposed partner cannot advance capital to the share he would like, he should take a smaller percentage of the project thereby making it possible to participate.

14.4 ADVANTAGES

Joint ventures spread the risk on larger projects. You can bid work beyond your individual bonding capacity. Sometimes larger jobs have fewer bidders, and it may be possible to get better prices for the work than on smaller jobs with a lot of bidders.

Estimates are more accurate as a result of the partners individually taking off quantities and pricing them. These are then compared for a good check as mentioned earlier. This is a tremendous advantage on large and complex projects. You are sure your estimate is accurate and can bid with less worry about adding a contingency for this part of the work.

You also have the potential to form a future joint venture with some of the partners on other projects where they would be the sponsor.

14.5 DISADVANTAGES

One big disadvantage is that you may educate a direct competitor in your methods and thinking. When you bid against him at a future time, he may know how to underbid you. Fortunately, this is rarely the case. You know who he is and should know his feeling in this matter ahead of time.

Profits will also be split in accordance with percentage of participation. When you are the sponsor and do the work in an efficient and profitable manner, others reap part of the rewards; but remember, they helped make the job biddable by adding their financial strength and participation during the construction process.

14.6 TYPES OF AGREEMENT

There are two basic types of agreement. The first is a pre-bid joint venture agreement. This is a short form which takes care of details and expenses up to bid day. It includes who bears the bidding expenses and the obligations of the partners including an agreement to enter into the joint venture if you are the successful bidder. The second, the post-award joint venture agreement, is the actual agreement under which the work is done. This part details the obligations of the partners and the percentage of participation and also includes how the venture will be financed. How the project will be manned and who will furnish the manpower needed is also stipulated. Management committees or sponsorship explaining the duties and rights of the parties, what develops in the case of a disagreement, how profits or losses are covered, and the remedies of the parties if one party does not perform as agreed are also outlined. Both of these agreements should be reviewed with your attorney and your C.P.A. There are legal obligations that you must understand when you agree to the joint venture. Your C.P.A. will advise you on the tax consequences, which are equally important to you. This post-award joint venture agreement should be set forth in outline form and agreed upon in advance of bidding the project.

This two-step arrangement eliminates the time consuming and potential costly legal efforts of making an agreement final when the bid is successful. The partners are aware of the conditions of the venture in advance, thereby eliminating any question as to what is expected of each of them.

Once it is determined that the joint venture bid is low, then the agreement has been worked out in final detail. Mutual trust and integrity are essential at this time.

Typical agreements are shown here for form only. Attorneys should be consulted for actual form and language.

14.6.1 Pre-Bid Joint Venture Agreement

VOLPE and (JOINT VENTURE PARTNER(S))

THIS JOINT VENTURE AGREEMENT, made and entered into as of the_____ day of_____ by and between VOLPE CONSTRUCTION, a Massachusetts corporation with offices at 54 Eastern Avenue, Malden, Massachusetts, hereinafter called "VOLPE" and_____, a corporation with principal offices at_____, hereinafter called _____ (hereinafter collectively called the "Joint Venturers"):

WITNESSED:

WHEREAS,

(Owner)   has asked for proposals for construction of the new   (Project Name) in   (*Location of Project*)   proposals expected to be received on   (Date Bid is due).

WHEREAS, the parties hereto have determined to estimate and bid jointly for the performance of said construction.

NOW, THEREFORE, in consideration of the premises and mutual covenants hereafter contained, it is hereby mutually agreed as follows:

1. The parties hereto do hereby join in joint venture for the purpose of bidding, negotiating and contracting for the construction of the said   (Project)   and supplements, change orders or amendments to said contract.

2. Each of the parties hereto shall make estimates of costs and in conference reach agreement as to conditions and prices of proposal to be submitted   (Owner) for the said construction.

3. All cost or expense incurred by either party hereto in estimating, interest, travel, or otherwise prior to the award of contract shall be borne and paid by the party incurring such cost or expense.

4. Each of the parties hereto shall designate and authorize individuals to be selected by them to represent and act for it in negotiating, arranging for prices and terms of contract, and entering into and to sign contract or contracts in behalf of the Joint Venturers as may be required by   (Owner)   , and each of the parties hereto agrees to furnish such power of attorney and certification of its Board of Directors as may be required to evidence and carry out said authorizations.

5. In the event that Joint Venturers are the successful bidder and are awarded contract for the said construction, they agree to enter into joint venture agreement further defining the interest of each of the parties hereto in and to the work to be performed under the said construction contract and any profits to be derived therefrom and any liabilities and/or losses incurred in connection thereof. (Tentative Agreement attached).

6. As between the Joint Venturers under this agreement and under the proposed joint venture agreements, their respective interests in profits or losses accruing from the

contract and project contemplated herein and obligations for contributions for joint venture working funds, and for disbursements and obligations in connection there with, shall be shared in the following proportion:

PARTICIPATION "VOLPE" _____%

" " _____%

7. In the event that performance bond, payment and material bond, or other bonds are required, each of the parties hereto does hereby agree to make such application and to furnish and execute such indemnity agreements as may be necessary to procure such bond or bonds, limited, however, to the respective percentages of participation as stated in Paragraph 6 above.

8. All necessary working capital when and as required for the performance of the construction contract shall be furnished promptly by the parties hereto proportionately in accordance with their respective percentages of participation as stated in Paragraph 6 above.

9. The parties hereto shall furnish and appoint "Project manager" and an "Office Manager" and shall furnish such assistance and available personnel as may be mutually agreed. Separate books of account and separate bank accounts shall be kept and maintained for the joint venture.

10. Neither party hereto shall sell, assign or in any manner transfer its interest or any part thereof in this joint venture without first obtaining the written consent of the other party.

11. This agreement and the relationship of the Joint Ventures shall be limited to the performance of the contract and the project contemplated herein. Nothing herein contained shall be construed to constitute the parties hereto as partners or as an association, legal or otherwise, or in any manner limit them in carrying on their respective businesses or activities in making other contracts or the performance of other work.

12. This agreement shall be subject to and interpreted under the laws of the (State where Project is located) or as mutually agreed.

13. This agreement and each of the terms and conditions hereof shall be binding upon and shall inure to the benefit of each of the parties hereto and their respective successors and assigns.

IN WITNESS WHEREOF, the parties hereto have caused these presents to be signed by their duky authorized officers and their repsective corporate seals to be hereunto affixed by like authority.

VOLPE CONSTRUCTION CO.

By _____

(JOINT VENTURE PARTNER(S)

By _____

14.6.2 Post-Award J V Agreement

This is a sample post award agreement:

JOINT VENTURE AGREEMENT made this   (date)   by and between

PARTIES   (names of the parties)

WITNESSED:

CONSIDERATION   That for good and valuable consideration paid by each of the parties hereto unto the other, the respective receipt whereof is hereby duly acknowledged, and in consideration further of the mutual promises, covenants and agreements herein contained,   and do severally represent and agree as follows:

PART I

BACK REFERENCES   A. Reference is made to   (Project name)   hereinafter referred to as the "PROJECT."

PART II

JOINT ACTION AND COOPERATION   A. VOLPE and   agree that they will cooperatively and jointly do and perform all or cause to be done and performed all acts necessary and required to complete the PROJECT subject to the terms and provisions of the herein agreement; in accordance with the procedures stated in the INVITATION and in the Letter of Acceptability, and otherwise as set forth in all applicable statues and regulations promulgated thereunder.

PERCENTAGE OF PARTICIPATION   B. As between VOLPE and   , their respective interests in profits or losses accruing from the PROJECT, and obligations for contributions for joint working funds and for disbursements and obligations in connection therewith shall be shared in the following proportion:

VOLPE   % participation
                      % participation

AND the parties do hereby agree to indemnify and save harmless the other against any loss or liability exceeding the proportions above stated whether resulting from payments required to be made by the execution of any surety company bonds or indemnity agreement in connection therewith, or from any payments arising out of any other indemnity agreement or guarantees or otherwise required to be made in the performance of the construction of the PROJECT.

BONDS

C. VOLPE and   agree to obtain and file a joint performance and payment bond in the penal sums and in the forms required by the contract documents and the joint venture agreement between the parties and in order to induce the surety company or companies to subscribe and file the performance bond and the payment bond—each in the sum required—the parties hereto do severally agree to indemnify the surety company or companies in amounts equal to the following percentages of the total bond penalties and limited to the following aggregate amount for each of the parties respectively:

VOLPE   % of any loss

                       % of any loss

JOINT BANK
ACCOUNTS

D. A joint bank account or accounts for the PROJECT shall be opened in such bank or banks and under such account name as may be agreed upon by the parties to the joint venture. Each party shall designate an authorized signatory or signatories on its behalf to endorse checks for deposit and to sign checks drawn on said account(s). All payments and money received from the owner or its mortgagee, shall be deposited in such account(s). All obligations of the parties in respect of the PROJECT shall be paid by check drawn on the said account(s).

WORKING FUNDS

E. Concurrently with the execution and delivery of this agreement and the opening of such bank account(s) VOLPE and   shall each pay into such joint account for initial joint working funds agreed upon as follows:

VOLPE $

                    $

ADDITIONAL

When and if the parties shall determine that the joint working fund is insufficient for the performance of the PROJECT, they shall pay over in their respective proportions, such additional amount(s) as may be required within five (5) working days thereafter. If either party should be unable or fail or neglect to contribute and deposit such additional funds in the joint bank account(s), then the other party shall have the right to advance the deficiency and upon making of such advance shall receive interest at the prime rate in effect at the time on such advance to the date of repayment; and such advance(s) shall be repaid in full with interest from the joint account(s) out of the first monies thereinafter received from the owner or its mortgagee and before any other payments are made from such joint account(s) to:

EXCESS WORKING FUNDS

F. If (prior to the completion of the PROJECT) the parties shall determine that the funds in the joint bank account(s) are in excess of the financial requirements of the PROJECT, such excess funds shall be distributed to VOLPE and   in accordance with their respective percentage participation and such distribution shall be deemed to be a pro rata return of working funds therefor advanced, until all such advances have been repaid in full; and thereafter such distribution shall be deemed to be PRO TANTO distribution of profit.

If after return of working funds there shall be a necessity for additional working funds (apart from or in addition to funds then, if any, on hand) the parties agree to make such monies available in their respective participation percentage and in accordance with the tenor of the foregoing portion of this section.

DISTRIBUTION OF FINAL COMPLETION

G. Upon final completion of the PROJECT, after the payment of all outstanding obligations, claims and indebtedness of the parties in respect of the PROJECT, or after sufficient reserves have been set aside therefor and for known or reasonably anticipated contingencies, any funds then remaining in the joint bank account(s) shall be distributed to VOLPE and   in accordance with their respective percentage participation. Thereafter, the retained balance, if any, of such reserves and contingency funds shall be similarly distributed as and when the purposes for such retention shall no longer exist. If performance of PROJECT shall result in a loss, then such loss (including any over-distribution of monies) shall be borne by VOLPE and   in accordance with their percentage participation.

BOOKS & RECORDS

H. Books of accounts and records for the PROJECT shall be kept and maintained under the supervision of an office manager. All books of accounts, records, vouchers, contracts and documents of any kind relating to the performance of the PROJECT shall be subject to examination by either party hereto at any time. Accounting methods shall be adopted as the parties shall prescribe.

COSTS

PROJECT cost shall include cost of construction, which shall include, without limiting the generality thereof, the cost of all subcontracts, bonds, the building permits, legal and accounting fees, losses and liabilities not covered or

compensated by insurance or otherwise incurred or suffered and arising out of or in connection with performance of the PROJECT, and such other costs and charges which by mutual agreement should be charged as a part of the cost of performance of the PROJECT. Cost of construction shall not include payment of any salaries or other compensation to officers, agent and employees of VOLPE or of   unless such salaries or compensation are specifically set forth and agreed upon in writing signed by VOLPE and
respectively; nor shall cost of construction include general overhead of VOLPE or of   directly or indirectly attributable to the PROJECT excepting as the same may be specifically agreed upon in writing and signed by the parties hereto. In order to reduce the administrative and accounting cost of the joint venture, either partner may enter into a lump sum subcontract with the joint venture to perform work including but not limited to concrete, masonry, general conditions, carpentry, and hand excavation.

AUDITING

From time to time, at written request of VOLPE or   and upon the completion of the PROJECT, the accounts shall be audited as an expense of the PROJECT by a firm of certified public accountants selected by the parties, whose audit and determination shall be final, conclusive and binding upon the joint venture partners.

GENERAL MANAGER

I. The PROJECT shall be carried out and performed under the direction of a General Manager to be selected and appointed by VOLPE and subject to approval of   .
The General Manager shall have full power of management and direction of the work and of the joint venture.

BOND AND
INSURANCE BROKER

J. Bonds and other coverage shall be placed by agreement of the parties.

DEPOSITORY

K.   (Name of Bank)   is designated as bank of deposit for the general and special accounts of the joint venture.

PART III

BOARD OF MANAGERS

A. Excepting as is specifically provided in PART II of this Agreement, the direction and control of the PROJECT shall be in a Board of Managers constituted and empowered as herein set forth. Said Board of Managers shall consist of a representative and an alternate designated by VOLPE and a representative and an alternate designated by   . Either party may, from time to time, change its representative or alternate but notice of such change shall be given in writing to the other party.

VOLPE does hereby designate as it's representative and as Alternate to serve in his absence or unavailability does hereby designate as it's representative and as Alternate to serve in his absence or unavailability does hereby designate

METHOD OF DECISION

The decisions of the Board of Managers shall be unanimous, but if they cannot agree upon any matter within their competence to decide, the decision of the representative (or alternate, as the case may be) of VOLPE shall control.

DISAGREEMENTS

In the event of such disagreement, the facts and issues in disagreement shall be reduced to writing by the aggrieved party and such grievance shall thereafter be handled in accordance with the grievance procedures hereafter established.

FUNCTIONS AND DUTIES OF BOARD

The representative (or alternate, as the case may be) of each party shall be vested with full power and authority to act for and in behalf of the party making such designation, in all matters involved in, or connected with or arising out of the subject PROJECT except where specifically provided for in PART II of this Agreement, to deal with the representative (or alternate, as the case may be) of the other party; to execute and deliver for and in behalf of the party making such designation any and all applications, agreements, documents, or other writing required by the construction contract to deal with all matters involving the interest of his corporation, VOLPE or , (as the case may be) in respect of the officers, structure and activity of this joint venture; to approve all subcontractors and subcontracts; to approve all purchases orders in excess of five thousand dollars and to approve the supplier; to approve all purchases or rental of tools and equipment; to arrange matters of banking and finances to the extent of the responsibility of his corporation VOLPE or , (as the case may be) therefore; and otherwise to represent the interest of the corporation making such designation in any matter not specifically provided for in this Agreement.

GRIEVANCE PROCEDURE

B. If, by reason of the inability of the Board of Managers to agree upon any matter within the scope, and (pursuant to the provisions of the foregoing Section A of this PART III) VOLPE shall make the decision in behalf of the Board of Managers; and if shall feel aggrieved thereby and shall duly file its grievance in writing with VOLPE who shall acknowledge receipt thereof.

ARBITRATION

If the grievance(s) shall not be earlier resolved to the satisfaction of (The aggrieved party) (and be thereby withdrawn) then any controversy which has so arisen under this contract and which might be the subject of a personal action at law or a suit in equity shall be resolved by the submission to the decision of three (3) arbitrators under the provisions of Section 14 through 22 of Chapter 251 of the General Laws of Massachusetts or its equivalent. One (1) arbitrator shall be chosen by each of the parties hereto and a third arbitrator shall be chosen by the two (2) so chosen. In case of death, inability or refusal to serve of any person so named, the Superior Court in and for the County of Suffolk shall upon application of either party hereto name an arbitrator or arbitrators, as the case may be. No person who is a party to this contract and no agent or agents, or employee or employees, of any party to this contract may be named as arbitrator. The submission of any controversy to arbitration shall be made within a reasonable time, and in all events within six (6) months, after due notice by either party claiming the arbitration of such controversy. The award of the majority of the arbitrators shall be made and reported to the Superior Court in and for the County of Suffolk within one (1) year from the date of submission, or within such further time as the Court may upon application of the Arbitrators allow; and the Judgement thereon shall be final. The costs of such arbitration shall be considered a PROJECT cost within the scope and meaning of Section I of PART II of this agreement, including legal fees.

PART IV

LIMITATION OF
HEREIN RELATIONSHIP
BETWEEN

A. The relationship between VOLPE and   shall be limited to the performance of the PROJECT under the terms of this agreement. Nothing herein contained shall be construed to constitute any relationship between VOLPE or   in the conduct of their respective businesses or activities, in the making of other contracts or the performance of other work, or impose any liability except that of performance of the terms, provisions and conditions of this Agreement.

The relationship created hereby shall terminate upon the completion of the PROJECT and the final closing thereof (including delayed performance of any "escrow" items) upon the distribution to the partners of all monies and property involved therein or other adjustment of the respective rights and obligations of the parties as in this agreement provided.

NON-ASSIGNABILITY   *B.* Neither this agreement nor any interest therein, or in the PROJECT, nor any interest in any monies belonging to or which may accrue to the PROJECT may be assigned, pledged, transferred or hypothecated except that in the event any party desires to obtain banking accommodations for the purposes of *this* agreement, such party may, with the prior written consent of all parties hereto, assign, pledge or hypothecate its right, title and interest in and to such monies as it may be entitled to receive hereunder to the lending bank as security for said banking accommodations.

THIRD PARTY RIGHTS   *C.* The right of any person, firm or corporation claiming by, through or under either party hereto (including but not limiting the same to judgment or other creditors, receivers, trustees, assignees, garnishees, executors, administrators, etc.) to exert any claim against the right, title and interest of any party thereto, shall be limited solely to the right to claim or receive after completion of the PROJECT and after the closing of the accounts between the parties, the distributive share of such party in the net proceeds payable hereunder, whether consisting of return of any contribution made to working funds hereunder, earnings or other assets, and then only subject to the equities of the other parties as in this agreement set forth.

INSOLVENCY   *D.* Should either party become bankrupt or insolvent or commit an act of bankruptcy, or take advantage of any bankruptcy reorganization, composition or arrangement statute, then such party (hereinafter called the "insolvent party") shall, from and after the happening of any of the foregoing, cease to have any participation in the management of the PROJECT; and its representation and the Board of Managers theretofore existing, shall be deemed cancelled and the power and authority previously possessed and exercised by the insolvent party shall be exercised solely by the other party. The insolvent party shall remain liable for its share of any losses, if any, and shall be entitled to receive its share of the profits, if any, all as provided in this agreement, to be paid at the times and in the manner provided in this agreement.

NAME   *E.* The relationship between VOLPE and
hereby in this agreement established and created shall be known as:

(Name of the Joint Venture)

NOTICES   *F.* Any notice that may be required under this agreement or any notice which—apart from requirement—either party hereto shall desire to give unto the other shall be sent by Certified Mail as follows:

TO: VOLPE

TO:

or acknowledge in writing by the signature of an officer of VOLPE or   , as the case may be, receiving the notice.

G. This agreement shall not be changed or modified except in writing and signed by each of the parties hereto.

H. This agreement shall be binding upon and shall insure to the benefit of only the parties hereto and their respective successors.

IN WITNESS WHEREOF, the parties hereto have herein to affixed their hands and seals the day and year first above written.

Volpe Construction Co.

By:

By:

14.7 CONCLUSION

My best advice is to joint venture only with known partners on projects of a type you and your people can handle. If it is a large complex project the partners should be of diverse experience thereby complementing each other. This will help make it a successful venture. A joint venture is not guaranteed to make money—it can also lose money.

CHAPTER 15

How Is Success Measured?

15.1 WHAT IS SUCCESS

Success is achieving a goal set by yourself; therefore, only you can be sure of the answer to this question.

The construction industry is very rewarding. You feel you have been a part of history when a building grows under your direction. And it does grow—from a hole in the ground to the finished product. It is challenging because every day you solve problems; some are routine and others are unusual, such as overcoming bad soil conditions or a tough concrete pour. In this industry you learn your shortcomings very fast. Be prepared to overcome them with competent help and methods. Construction does not give you the opportunity to make many mistakes—they are too costly.

15.2 RECORDS AND RELATIONSHIPS

For success, records must be kept up-to-date. As in all phases of our life, we learn from experience. From past projects we learn how to bid on future work whether it be what we have done before or more complicated.

Record keeping ought to be tailored to your own needs. The sample forms in this book have contributed to the success of our firm and possibly they can help you. These forms can be modified for your specific needs if you so desire.

Records tell you what type of project you handle with the greatest efficiency, which ones are most profitable, and even which clients you are more compatible with.

It is advantageous to have a brochure showing the projects you have completed. It should give the names of owners and architects with whom you have done business so that those who read the brochure can call for references. It has been said that one picture is worth a thousand words, and this is why a good brochure is needed.

Make your organization efficient in every area of your practice. It does not have to be big, but it should be competent.

A tremendous feeling of satisfaction comes from the relationships you make, keep, and nurture with your clients, subcontractors, labor, and other contractors. These relationships are important in helping develop your business.

15.3 SERVICE TO OTHERS

Public service in civic endeavors and association work is rewarding. Your ideas for improvements can be promoted and you will have the satisfaction of seeing some come into reality.

Through continual vigilance, education, and association work, you will get to know about industry problems and how to handle them. When you are aware of the problems and what the consequences will be, then you certainly can and will act more prudently. This is our industry, and much hard work is needed to keep it an honorable and rewarding one. These rewards include both the feeling of accomplishment and monetary gain.

15.4 BE READY FOR CHANGE

You must be ever on the alert for change, for new contracting methods, construction methods, and materials. You have to learn how to handle and work with these new methods, systems, and materials. It is not unusual to see an old-line firm get into difficulties because they have refused to accept these changes.

Construction is always changing, and unless you improve your relationships and methods, the whole system can change without you. Owners continually become more sophisticated and demanding. Because of higher interest rates for money, they insist on shorter construction time and better results. You must be able to accommodate them or others will be eager to do the job.

Computerization has come of age. It seems we cannot do anything without a computer. Today most classrooms in all grades have them. Owners and architects use them and so do we. All facets are worked out on the computer—feasibility studies, time schedules, project schedules, budgets, estimates, cost reports, payment records, and on and on. Do not overlook this important tool.

Many of us know what the industry is trying to accomplish, but do we convey this clearly to those outside our circle of interest?

Our purpose in writing this book is that it may stimulate your thinking, aid you in becoming more successful, and give you a better understanding of the construction business and obligations of the general contractor and construction manager, in whatever type of construction you do.

Some Building and Engineering Construction Documents and A.G.C.A. Supervisory Training Services

Subcontract Form of Contract

Project Administrator Form of Contract

Building Construction Documents

DESIGN-BUILD DOCUMENTS

Preliminary Design-Build Agreement. This agreement provides for reimbursement of design expenses in the event the proposed project does not go forward. 1980.
Order No. 400

Standard Form of Design-Build Agreement and General Conditions Between Owner and Contractor (Provides a Guaranteed Maximum Price). This agreement spells out the terms of a formal agreement for design-build services. 13 pages. 1982.
Order No. 410

Standard Form of Design-Build Agreement and General Conditions Between Owner and Contractor (Where the Basis of Compensation is a Lump Sum). This agreement is similar to the 410 but is based on lump sum compensation. 11 pages. 1986.
Order No. 415

Conditions Between Contractor and Subcontractor for Design-Build. This agreement is a companion to 410, 450 and 450-1. 20 pages. 1982.
Order No. 430

Standard Design-Build Subcontract Agreement With Subcontractor Not Providing Design. This agreement is designed to be used with AGC Document 430. *Conditions Between Contractor and Subcontractor for Design-Build*, where the subcontractor does *not* perform design functions. 1983.
Order No. 450

Standard Design-Build Subcontract Agreement With Subcontractor Providing Design. This agreement is designed to be used with AGC Document No. 430, *Conditions Between Contractor and Subcontractor for Design-Build*, where the subcontractor provides design services. 1983.
   Order No. 450-1

Standard Form of Agreement Between Contractor and Architect. This document was developed by AGC's Special Contracting Methods Committee for use by contractors engaged in the design-build method of contracting when an architect is used to provide the design services. 10 pages. 1985.
   Order No. 420

Change Order/Contractor Fee Adjustment. Developed by AGC's Special Contracting Methods Committee for use by contractors in handling change orders on design-build projects or on any project in which a fee is involved.
   Order No. 440

Design-Build Guidelines. This small brochure explains the advantages of design-build contracting and how to select a design-build contractor. A project owner aid. 5 pages. 1977.
   Order No. 405

CONSTRUCTION MANAGEMENT DOCUMENTS

Standard Form of Agreement Between Owner and Construction Manager. This contract allows the construction manager to do work with its own forces and provides a guaranteed maximum price option. 15 pages. 1980.
   Order No. 500

Amendment to Owner-Construction Manager Contract. This agreement makes provision for exercising the guaranteed maximum price option of AGC No. 500. 1977.
   Order No. 501

Standard Form of Agreement Between Owner and Construction Manager (Owner Awards All Trade Contracts). This contract may be used with the CM process when the owner awards all the trade contracts. 11 pages. 1979.
   Order No. 510

General Conditions for Trade Contractors Under Construction Management Agreements. This contract is the companion to AGC No. 500 & 510. 20 pages. 1980.
   Order No. 520

Change Order/Construction Manager Fee Adjustment. Developed by AGC's Construction Management Committee for use with projects being built under the construction management method of contracting.
Order No. 525

Construction Management Guidelines. These guidelines give AGC's position on the proper role of a construction manager and each member of the "team." 13 pages. 1979.
Order No. 540

Construction Management Control Process. These guidelines were developed to answer the need of the construction industry for an outline of the normal process to be followed on a construction management project. This booklet charts and describes the responsibilities of the owner, architect, and construction manager on a construction management project. 6 pages. 1976.
Order No. 545

Owner Guidelines for Selection of a Construction Manager. This publication provides several suggested points on what qualifications a construction manager should have, as well as providing guidelines for the selection procedures. 8 pages. 1982.
Order No. 550

Construction Management Delivery Systems for Hospital Facilities. "Construction Management (CM) Delivery Systems for Hospital Facilities—5 Case Studies"—proceedings from the 1983 AGC National Symposium. Learn how hospital owners, construction managers, and architects/engineers used the "Team Approach" to build five different hospital projects within the optimum time, for the most economical cost, and the required quality through construction management (CM). 1983.
Order No. 560

Construction Management Kit. This publication is a compilation of the work of AGC's Construction Management Committee over the past 16 years. It includes CM contract documents, guideline information for the CM, and other publications pertaining to construction management services. The documents and publications in the Kit express "state of the art" concepts on the construction management method of construction contracting. 1988.
Order No. 590

SUBCONTRACT DOCUMENTS

Standard Subcontract Agreement For Building Construction. The Subcontract For Building Construction is a state-of-the-art contract form, incorporating the

conditions prevalent on a construction project today. The 15-page subcontract includes 16 articles which cover in detail such provisions as scope of work, schedule
of work, contract price, payment, contractor's and subcontractor's obligations,
indemnification, insurance, arbitration, and contract interpretation. The subcontract
also includes sections which allow the user to incorporate provisions specific to the
subcontract, such as the scope of work, common temporary services, insurance
coverage amounts, and special provisions. Appropriate labor relations provisions also
may be specifically tailored and inserted in the subcontract. 1984.
Order No. 600

Short Form Subcontract. This convenient form is specifically designed for use on
subcontract work that is of limited dollar value and will be completed within a
relatively short time. The form has instructions for all fill-in provisions, and had just
nine general articles. 4 pages. 1987.
Order No. 603

Subcontract For Use On Federal Construction. This subcontract is intended
to be used on direct federal building or engineered construction projects. It reflects
the relevant provisions of the Prompt Payment Act Amendments of 1988 and the
Contract Disputes Act, and references the Federal Acquistion Regulation.
Order No. 601

Standard Subbid Proposal. This form is for subcontractors to describe the scope
of work covered in their bids. Developed as a guide by AGC, the National Electrical
Contractors Association, The Mechanical Contractors Association of America, the
Sheet Metal and Air Conditioning Contractors National Association, and the National
Association of Plumbing-Heating-Cooling Contractors.
Order No. 605

Subcontract Performance Bond. This bond form can be requested by a contractor from a subcontractor to guarantee the subcontractor's performance. This form
was developed with the assistance of the Surety Association of America. 1988.
Order No. 606

Subcontract Payment Bond. This bond form can be requested by a contractor
from a subcontractor to guarantee that the subcontractor will pay laborers and material suppliers. This form was developed with the assistance of the Surety Association of America. 1988.
Order No. 607

Subcontractor's Application for Payment. This application provides a standardized format for subcontractor's requests for payment.
Order No. 610

Invitation to Bid Form for Subcontractors. This form is an invitation to a subcontractor to submit a bid on a specific project.
   Order No. 614

AIA CONTRACT DOCUMENTS AND AGC PUBLICATIONS

General Conditions of the Contract For Construction, 1987 Edition. Fourteenth Edition. AIA Document A201, the basic contract document used by many owners and architects on building construction. 24 pages. 1987.
   Order No. 301

Standard Form of Agreement Between Owner and Contractor-Stipulated Sum. This is AIA Document A101. Twelfth Edition. The basis of payment is a stipulated sum (fixed price). 8 pages. 1987.
   Order No. 300

Abbreviated Form of Agreement Between Owner and Contractor-Stipulated Sum. This is AIA Document A107 and is inteded for use by parties who have an established course of dealing on other projects. It contains its own general conditions, stipulated sum. 11 pages. 1987.
   Order No. 305

Standard Form of Agreement Between Owner and Contractor—Cost Plus Fee. This is AIA Document A111 for use when the contract is for the cost of the work plus a fee. 14 pages. 1987.
   Order No. 310

Abbreviated Form of Agreement Between Owner and Contractor—Cost Plus Fee. This is AIA's A117 which combines agreement clauses and general conditions provisions. It is designed for use on projects of limited scope where the basis of payment is cost of the work plus a fee. It also contains an optional GMP. 13 pages. 1987.
   Order No. 317

Contractor's Qualification Statement. This document is AIA Form A305. It is used by contractors to indicate the experience of their firm and principal officers. It also is used to assist architects and owners in evaluating the contractor. 1986.
   Order No. 320

Instructions to Bidders, 1987 Edition. This is AIA Form A701. It contains complementary provisions and is designed to be used with the 1987 AIA A201. 5 pages. 1987.
   Order No. 321

Recommended Guide for Competitive Bidding Procedures and Contract Awards For Building Construction. This guide, AIA Document A501, is published by AIA and AGC. It is useful in presenting bids to owners or architects. A must for building contractors. 7 pages. 1982.
  Order No. 325

Guide to Supplemental Conditions. This is AiA Form A511. This guide points out the kinds of additional information most frequently required to cover local situations and variations in project requirements. Even though it suggests standardized language, it is not meant to be a standardized form of supplementary conditions. Rather, it is inteded to be an aid to the practitioner in preparing supplementary conditions. 1987.
  Order No. 327

Change Order. This change order is AIA Form G701 and is used by architects to order contractors to perform changes in the work. 1987.
  Order No. 330

Application and Certificate for Payment and Continuation Sheet. This application for payment is comprised of AIA Forms G702 and submitted by the contractor to the owner through the architect. May 1983.
  Order No. 335

Continuation Sheet. This contiuation sheet is comprised of AIA Form G703 and is submitted by the contractor to the owner through the architect. May 1983.
  Order No. 336

Certificate of Substantial Completion. This is AIA Form G704 and is used to certify that the project is substantially complete. April 1978.
  Order No. 340

Comparisons of B141 and A201 1976/77 and 1987 Editions. This workbook is a side-by-side reprint of the comparison of the changes in AIA's two main documents in the 1987 edition: changes in B141, Standard Form of Agreement Between Owner and Architect, from the 1977 edition, and changes in A201, General Conditions of the Contract for Construction, from the 1976 edition. There are editorial features that highlight the changes in the documents. The document is designed to help users make and informed transition to use of the 1987 AIA documents. 1987.
  Order No. 345

STANDARD CONTRACTS AND CONTRACT DOCUMENTS

Standard General Conditions of the Construction Contract (for Engineering Construction). This is the Document that constitutes the basic agreement between the owner and the contractor. It defines the terms used in contract documents and sets forth in detail the basics of the duties and responsibilities of the contractor and owner and all of the provisions basic to the contract for construction. (EJDCD 1910-8.) 1983 Edition.

Order No. 200

Contract Documents for Construction of Federally Assisted Water and Sewer Projects. These AGC endorsed contract documents are acceptable for use by borrowers and grantees in federally assisted projects funded by EDZ, FHMA, EPA, and HUD. Comprised of eleven key documents, they are also endorsed by the three major engineering societies and APWA. 36 pages. 1983 Edition.

Order No. 201

Standard Form of Agreement Between Owner and Contractor On the Basis of a Stipulated Price. These documents provide compensation provisions and conditions for either of the stated basis. They deal with contract time, price, cost of work, contractor's fee, guaranteed maximum price, changes in the work, payment procedures and various administrative provision. (EJCDC 1910-8-A-1.) 1983 Edition.

Order No. 202

Standard Form of Agreement Between Owner and Contractor On the Basis of Cost-Plus. See above desciption. (EJCDC 1910-8-A-2.) 1983 Edition.

Order No. 203

Subcontract for Use on Federal Construction. This subcontract is intended to be used on direct federal building or engineered construction projects. It reflects the relevant provisions of the Prompt Payment Act Amendments of 1988 and the Contract Disputes Act, and references the Federal Acquisition Regulation.

Order No. 601

Change Order. This document provides a format for intitating a change in the work including price, time or both. (EJCDC 19190-8-B). 1983 Edition.

Order No. 204

Certificate of Substantial Completion. This form indicates acceptance of work to the extent stated and is to be accompanied by a "punch list" of items to be completed and spells out owner and contractor responsibilities prior to final payment. (EJCDC 1910-8-D). 1983 Edition.

Order No. 205

Application for Payment. This form provides space to spell out value of work involved and quantity and amount of work completed together with payment due. (EDCDC 1910-8-E). 1983 Edition.

Order No. 206

Work Directive Change. Form is to be used for changes intitiated in the field which may be included in subsequent change orders. (EJCDC 1910-8-F). 1983 Edition.

Order No. 207

Engineer's Letter to Owner Requesting Instructions Re: Bonds and Insurance During Construction. Protype letter to owner requesting instructions on types of coverages and amounts of insurance to be stated in contract documents. (EJCDC 1910-20). 1983 Edition.

Order No. 211

Owner's Instructions to Engineer Re: Bonds and Insurance During Construction. Sets forth the terms and conditions of performance, labor and material bonds to be provided by the contractor. (EJCDC 1910-21). 1983 Edition.

Order No. 212

Supervisory Training Services

HOW STP WORKS

STP consists of eleven units emphasizing leadership, motivation, communications, cost awareness, planning and scheduling—the skills supervisors must have to be effective. Each of the ten STP-Superintendent units are 20–25 hours of intensive classroom training. STP-Foreman is a 40-hour overview course.

Participants in the STP courses draw upon their field experience and learn by interaction with other students from all areas of the construction industry.

STP courses are sponsored by AGC chapters, construction firms, community colleges, adult education groups, vocational-technical education institutions, and labor groups, with highly qualified instructors following prescribed, proven procedures for STP instruction. There are more than 100 prime sponsors of STP courses in 48 states, who have offered more than 1.5 million contact work-hours of STP classroom instruction over the past thirteen years.

STP is constantly being revised to keep abreast of supervisors' needs and the opportunity to improve productivity in the industry.

SPECIAL STP SERVICES

STP Instructor Training Workshops are offered annually, where potential instructors learn about the STP philosophy, getting the maximum participation from STP enrollees, and teaching tips to help make the STP courses enjoyable and beneficial. The workshops feature the STP Instructor's Handbook as a guide.

To help new sponsors establish local STP programs, the STP Consulting Service with recognized STP experts will come to your area at your convenience to assist you.

Local promotion of STP programs is easy with *The STP Story,* (Order No. 28) a 10-minute videotape presentation on the benefits of STP training. Included is information on how to sponsor STP courses, choosing the right instructor, classroom teaching tips, and testimonials on what STP training can do for the individual and the employer.

STP participants need recognition for their achievements . . . and the STP Hardhat Recognition Decals are the right way to reward this dedication. There is one sticker for each of the eleven STP units, which is awarded with the unit completion certificate on the last night of class. For those who complete eight, nine or ten of the STP-Superintendent units, there are completion certificates and wallet cards to highlight this special achievement.

STP maunuals are available for purchase directly from Wil McKnight Associates, Inc. An order form appears on page 8. Please allow four to six weeks for book delivery.

The STP Story. A ten-minute video presentation on the benefits of STP training. Included is information on how to sponsor STP courses, choosing the right instructor, classroom teaching tips, and testimonials on what STP training can do for the individual and the employer.
Order No. 28
Call the AGC Education and Research Foundation to schedule a rental period.

Supervisory Training Program Wall Chart. This STP foldout wall chart (22" by 32") provides a session-by-session description of the program content of the 10 superintendent courses and the foreman course. For more information on STP, please refer to page 5, 1987.

Order No. 25

Supervisory Training Program Instructor's Handbook. A guidebook with helpful hints on successfully organizing and teaching local STP courses. Highlights important STP instructional methods and instructor skill development (including class discussions, lectures, and role playing). See page 5 for more details on the STP Program. 1987.

Order No. 26

STP Catalog. An 8-page brochure outlining the content of each of the 10 STP-Superintendent units and STP-Foremen. Describes special STP services available to local sponsors. Perfect for promoting local upcoming STP course offerings. 1987.

Order No. 27

The Volpe Construction Co., Inc. Subcontract Agreement

THIS AGREEMENT, made this   day of
by and between

                                                            , hereinafter called the Subcontractor
and THE VOLPE CONSTRUCTION CO., INC., hereinafter called the Contractor.

WITNESSETH, That the Subcontractor and Contractor agree as follows:

Article 1.   The Subcontractor shall furnish and pay for all materials, plant, equipment, tools, scaffolds, and shall furnish all labor (regardless of trade jurisdiction), and services, and do all other things necessary to fully execute the following work:

hereinafter called the Work, for the construction of

                                                            , hereinafter called the Project,

for
hereinafter called the Owner, at

subject to and in strict accordance with the Plans, Specifications, General Conditions, Supplemental General Conditions, Special Conditions and Addenda No.(s)
prepared by
hereinafter called the Architect, and with the terms and provisions of the General Contract, between the Contractor and the Owner dated   hereinafter called the General Contract,
and with the Additional Provisions, pages(s)   annexed

hereto and made a part hereof. All documents hereinabove mentioned, hereinafter referred to as the Contract Documents, become a part of this Agreement and are available for examination by the Subcontractor at the Contractor's office.

Article 2. The Subcontractor shall complete the several portions and the whole of the Work in accordance with the Project Schedule and as directed by the Contractor, so that the Project can be completed by as required by the General Contract.

The Subcontractor accepts the obligation of scheduling and coordinating its own work in conformance with the overall project schedule and shall adjust its schedule and coordinate its work with the work of others as directed by the Contractor.

The Subcontractor fully understands that any delay on its part in the performance of the Work will delay the Work of the Contractor and other Subcontractors and to this end agrees to provide sufficient quantities of skilled workmen, materials and equipment to begin, prosecute and complete the Work in a diligent and orderly manner and without interruption so as to insure the completion of the Work and the Project in accordance with the Contract Documents.

Should the Work or the Project be delayed by any fault or neglect or act or failure to act of the Subcontractor, the Contractor reserves the right to order the Subcontractor to work overtime, at the Subcontractor's expense, to make up for all time lost and to avoid delay in completion of the Work and of the Project.

Should the Subcontractor continue to fail to keep up with the schedule of the Work or the progress of the Project, or otherwise fail to comply with the terms and provisions of this Agreement, the Contractor, after three (3) days written notice to the Subcontractor, reserves the right to purchase materials, tools and equipment and employ labor and to do or cause to be done, all or any portion of the Work, by whatever means or methods the Contractor may deem expedient, all at the expense of the Subcontractor.

The Subcontractor agrees that upon receipt of notice, verbal or written, that the Subcontractor is not prosecuting the work or is otherwise failing to comply with the terms and provisions of this Agreement it will not remove any of its equipment, tools or materials from the Project until written permission has been received from the Contractor.

It is hereby mutually agreed by and between the Contractor and the Subcontractor that the continued progress of the work is an essential condition of this Agreement, and it shall be considered that time is of the essence in this Agreement.

Article 3. The Contractor agrees to pay the Subcontractor for the satisfactory performance and completion of the Work and of all the requirements of this Agreement, the total sum of

. . . .

. . . . Dollars ($

hereinafter called the Price.

The Price includes all labor, materials, plant, staging, scaffolding, use of equipment or rental of equipment, tools, benefits due labor under Union contracts, insurance premiums, along with all permits, fees, inspection costs and taxes, inclusive of sales tax, in connection with the Work required by the Contract Documents, or otherwise required by any Public Authority or by any Federal, State, County or Municipal Government or subdivision thereof. The Price is subject only to additions and deductions for changes as may be authorized by the Contractor.

Payments to the Subcontractor are to be made monthly within 5 days of the receipt of payment by the Contractor from the Owner for 90% of work performed or materials incorporated by the Subcontractor and accepted on the project the preceding month.

As a condition precedent to any partial payment, the Subcontractor shall submit such evidence and substantiation as the Contractor may require, showing that all obligations and payments due sub-subcontractors, materialmen and others in connection with the Work have been made by the Subcontractor when due.

Final payment shall be due 45 days after receipt of final payment from the Owner by the Contractor, provided, first, however, that (1) the Subcontractor shall have furnished evidence satisfactory to the Contractor that there are no claims, obligations or liens of sub-subcontractors, materialmen, or others outstanding in connection with the Work, and (2) the Subcontractor shall have executed a General Release and Indemnification Agreement, in form satisfactory to the Contractor, in favor of the Contractor and the Owner. (3) The Contractor shall have received final payment from the Owner.

Article 4. With respect to the work to be performed and furnished by the Subcontractor hereunder, the Subcontractor agrees:

4. (a) to be bound to the Contractor by all of the terms and provisions of the General Contract and the other Contract Documents and to assume toward the Contractor all the duties, obligations and responsibilities that the Contractor by those Contract Documents assumes toward the Owner.

4. (b) that an extension of time for the completion of this Agreement is hereby granted to the Subcontractor for a period equal to any delay caused by the Contractor. Such extension of time shall be in lieu of and in full satisfaction of any and all claims whatsoever of the Subcontractor against the Contractor.

The foregoing is not applicable to any act or omission or interference of or by the Owner which causes delay or damage to the Subcontractor. Such claim resulting therefrom shall be submitted by the Subcontractor under the terms of and in accordance with the requirements of the Contract Documents.

4. (c) that the Contractor is not an insurer or guarantor of the Work of the Contract Documents or of the performance of the Owner under the General Contract as specified therein or otherwise. The Subcontractor shall be bound by any changes, additions, and/or deletions to the Work ordered by the Owner or under the terms and provisions of the Contract Documents to the same extent that the Contractor is bound thereby. Where changes are ordered or where claims for extra work are made by the Subcontractor, the Subcontractor shall submit the dollar value of the work to be changed, added and/or deleted in the manner provided in the Contract Documents.

Should the parties hereto be unable to agree to the value of the work to be changed, added or deleted, the Subcontractor shall proceed with the work without delay provided the Contractor issues a written order from which order the value of the changed work will be omitted, and agrees to refer the matter under dispute to the Architect for a final decision.

If the Contract Documents authorize an appeal from the act or decision of the Architect or Owner and the time for such appeal has not expired, the Contractor, upon written request from the Subcontractor, may at its sole option, either prosecute an appeal on behalf of the Subcontractor under the provisions of the Contract Documents, or permit the Subcontractor to prosecute an appeal in the name of the Contractor, either option to be at the sole expense of the Subcontractor, including legal fees and any other costs of litigation.

All changes, additions or deletions in the Work ordered in writing by the Contractor are deemed to be a part of the Work of this Agreement.

4. (d) that the Contractor will not be responsible for any loss or damage to the work covered by this Agreement, however caused, including loss of or damage to materials, tools, equipment and personal property owned, rented or used by the Subcontractor, its agents, servants, employees, sub-subcontractors or materialmen.

4. (e) that fire insurance (with extended coverage, if specified) will be provided by others only for materials and equipment incorporated in the Project and all materials and equipment on the Premises intended for incorporation in the Project, in accordance with and to the limits listed in the Specifications.

4. (f) to hereby assume the entire responsibility and liability for any breach of the provisions of this Agreement and for any and all damages or injuries of any kind or nature whatever (including death resulting therefrom) to all persons, whether employees or otherwise, and to all property arising out of or occurring in any way connected with the execution of the work. The Subcontractor agrees to indemnify and save harmless the Contractor, its officers, stockholders, agents and employees from and against any and all claims, suits, losses, damages, judgments, costs, expenses and charges of every kind and nature, both legal and otherwise, arising out of or occurring in any way connected with the execution of the Work or arising out of or in any way connected with any breach of the Subcontractor of the provisions of this Agreement, whether the same arise under Workmen's Compensation Law or Common Law or otherwise. If any such claim or demand is made or threatened against the Contractor, its officers, stockholders, agents or employees, the Subcontractor agrees to assume and pay for the defense of any action at law or in equity resulting therefrom and to pay any judgment resulting from such action. The Contractor shall have the right to withhold from any payment due or to become due the Subcontractor an amount sufficient to cover any potential or actual loss and expense, including but not limited to judgments, legal fees and other costs of litigation, unless the Subcontractor shall furnish a surety bond, satisfactory to the Contractor guaranteeing such protection in accordance with the foregoing. Before commencing the Work the Subcontractor shall procure, carry and maintain at its own expense, for the full duration of the Work, all insurances in accordance with requirements of the Specifications and at least to the limits indicated in the Additional Provisions entitled Insurance Requirements.

4. (g) if the Subcontractor is allowed to use or uses any equipment, tools, or facilities, manned or unmanned, gratuitously or otherwise, which are owned, leased, or rented by the Contractor, the following terms and conditions

shall apply: The Subcontractor agrees that the equipment and all persons operating such equipment, including the Contractor's employees, are under the Subcontractor's exclusive jurisdiction, supervision and control, and the Subcontractor agrees to indemnify and save the Contractor, its employees and agents harmless from all claims for death or injury to persons, including the Contractor's employees, and from all loss, damage, or injury to property, including the equipment, tools, or facilities, arising in any manner out of the Subcontractor's operation or use of such equipment, tools, or facilities, whether or not based upon the condition thereof or any alleged negligence of the Contractor in permitting the use thereof. The Subcontractor's duty to indemnify hereunder shall include all costs or expenses arising out of all claims specified herein, including all court and/or arbitration costs, filing fees, attorney's fees and costs of settlement. The Subcontractor agrees to provide competent and experienced personnel to direct the operation of the equipment at all times, and further agrees that the Standard Crane and Derrick Signals in accordance with American Standard B 30.5-1968 shall be used to direct the equipment at all times when applicable. The Subcontractor further agrees to use said equipment in accordance with the manufacturer's instructions and agrees not to exceed the manufacturer's load capacities for such or similar equipment. The Subcontractor expressly agrees that counterweight in excess of the manufacturer's specifications shall not be used. The Subcontractor agrees to use the equipment in strict compliance with all applicable rules, laws, regulations, and orders. The Subcontractor assumes all liability for the adequacy of design or the strength of any sling, rope or strap and of any lifting lug or device embedded in or attached to any object.

4. (h) at its own expense to conform to the basic safety policy of the Contractor, and comply with specific safety requirements promulgated by any governmental body or authority, including but not limited to the requirements of the Occupational Safety and Health Act of 1970 and the Construction Safety Act of 1969 and all standards and regulations which have been or shall be promulgated by the parties or agencies which administer such Acts. Subcontractor shall have and exercise full responsibility for compliance hereunder by its agents, employees, materialmen, and Subcontractors generally, and in particular with respect to its portion of the Work on the Project and shall itself comply with said requirements, standards and regulations, and require and be directly responsible for compliance therewith on the part of its said agents, employees, materialmen and sub-subcontractors. The Subcontractor shall directly receive, respond to, defend, and be responsible for all citations, assessments, fines or penalties which may be incurred by reason of the Subcontractor's failure or failure on the part of its agents, employees, materialmen or sub-subcontractors to so comply.

4. (i) that the Subcontractor will be responsible for and provide fire protection for his own specialty requirements.

4. (j) that the Subcontractor shall (1) keep the premises free at all times from debris and waste materials caused in connection with the Work, and in particular leave each area of the Work in a condition satisfactory to the Contractor so as to be ready for succeeding work; (2) clean from its own Work and that of others any soiling or staining caused by the execution of its Work and repair any defects caused therefrom.

The Contractor will be responsible for removal from the premises of debris and waste materials deposited by the Subcontractor in areas and/or containers as designated by the Contractor.

Should the Subcontractor repeatedly fail to diligently comply with the foregoing, the Contractor shall have the right to perform or cause the clean-up work to be performed at the Subcontractor's expense. If the Subcontractor is one of a group that is not performing its clean-up work, the costs of clean-up will be divided among the group on a pro-rata basis in accordance with the best judgment of the Contractor.

4. (k) to make all claims, for any reason and of whatever nature, to the Contractor in the manner provided in the Contract Documents for like claims by the Contractor upon the Owner. The Subcontractor agrees that the total amount to be paid to the Subcontractor by the Contractor for such claims is the amount paid to the Contractor for the Subcontractor's account by the Owner for such claim. The Subcontractor agrees that if he does not comply with the foregoing all such claims are considered waived and any claim or suit brought thereon is invalid and without any basis in law or in fact.

4. (l) to diligently repair or remove and replace in accordance with the direction of the Architect or the Contractor any of the Work, materials, or equipment which the Architect or the Contractor shall condemn as defective or failing to conform to the requirements of the Contract Documents and to be fully responsible for any damage to other work caused as a result of such defective work.

Article 5.

5. (a) The Subcontractor agrees not to employ men, means or methods that may cause disruption of the Work, or of the work of others. In the event of a jurisdictional dispute involving the Subcontractors Work for this project, whether performed on site or off site, the Subcontractor agrees during the lifetime of this Agreement, to be bound to the provisions and procedural rules of the plan for the Settlement of Jurisdictional Disputes in the

construction industry. In the event of a dispute, the Subcontractor hereby agrees that he will immediately take the necessary action to resolve this dispute through said board so as to minimize delays or additional costs to other parties involved on this project. It is fully understood and so agreed that any decision or interpretation by said boards (or hearings panel) shall immediately be accepted and complied with. Subcontractor further agrees that in the event he subcontracts any part of the work covered by this subcontract, such additional subcontractor shall be required as a part of this subcontract to settle all jurisdictional disputes on this project in accordance with the procedures set forth in this paragraph.

5. (b) The Subcontractor agrees that the Contractor is not in any way or for any reason bound by any agreement that may exist or be made during the course of this Agreement between the Subcontractor and any individual, group or organization.

Article 6. If the Subcontractor should become insolvent or be adjudged a bankrupt, or make a general assignment for the benefit of his creditors, or otherwise acknowledge insolvency, or have a Receiver appointed, or if he should refuse or should neglect to supply enough properly skilled workmen or proper materials, or if he should fail to make prompt payment to Sub-subcontractors or for material or labor, or otherwise be guilty of a violation of any provision of this Agreement or the Contract Documents, or should he disregard Federal, State, County or municipal laws, ordinances or regulations, or the instructions of the Contractor, then the Contractor may without prejudice to any other right or remedy of the Contractor and after giving the Subcontractor and his surety, if any, three days' written notice, terminate the employment of the Subcontractor for all or any portion of the Work, and enter upon the premises of the Subcontractor and on the site and take possession of the premises and all materials, equipment, tools, appliances and any other items thereon or on the site, all of which the Subcontractor does hereby transfer and assign to the Contractor for the purpose of completing the Work and complete the Work by whatever means or methods the Contractor may deem expedient.

In case of such termination of the Subcontractor, the Subcontractor shall not be entitled to receive any further payment until the Work has been completed and accepted by the Owner at which time if the unpaid balance of the Contract Price shall exceed such costs and expense incurred by the Contractor to complete the Work, such excess shall be paid to the Subcontractor; but if such cost and expense exceed the unpaid balance, the Subcontractor shall pay such difference to the Contractor. Such cost and expense shall include, in addition to the cost of completing the work, all losses, damages, costs and expenses, including legal fees and other costs of litigation, managerial and administrative services incurred by reason of or resulting from the Subcontractor's default.

Article 7. In the event of a dispute arising under this Agreement or the Contract Documents, which is not disposed of by agreement between the parties hereto, the Subcontractor, upon written notice from the Contractor, agrees to proceed diligently with the performance of the Work so as not to delay or affect the completion of the Work or the work of others. It is agreed that by proceeding with the Work the Subcontractor has not forfeited any of his rights under this Agreement or his rights under the Contract Documents.

Article 8. The Subcontractor agrees that no mechanics liens or claims will be filed or maintained against the Project or any part thereof, or against any monies due or to become due from any agency to the Owner or from the Owner to the Contractor or from the Contractor to the Subcontractor, and in the event a lien or claim is filed in connection with the Work of this Agreement the Subcontractor will cause such liens and claims to be satisfied or removed at its own expense within 7 days from the date the Subcontractor is notified of such filing and failing to do so the Contractor shall have the right to cause such liens or claims to be satisfied or removed by whatever means the Contractor chooses. The Subcontractor agrees to be responsible for all costs and expenses, including attorney fees and other costs of litigation, and to protect and save harmless the Contractor and the Owner from and against any damages and liability in connection with the satisfactory removal of such liens or claims.

Article 9. The Subcontractor agrees, that in the event that arbitration as set forth within the provisions of the Contract Documents is claimed against the Contractor in connection with the Work of this Agreement, either by the Owner, another subcontractor, or any other person or entity, and arbitration is in turn claimed by the Contractor against the Subcontractor, at the request of the Contractor, the Subcontractor will consent to a consolidation of the arbitration and defend and save harmless the Contractor against all damages, liability, cost and expenses, including legal fees and other costs of litigation which the Contractor may sustain in connection therewith.

Article 10. The Subcontractor hereby guarantees the Work to the full extent required by law or by the Contract Documents or if no guarantee is called for agrees to remove, replace, and/or repair, at its own expense, any defective or improper work, material or equipment discovered within one year from the date of the acceptance of the Project as a whole by the Owner.

The Subcontractor, in addition to the foregoing, agrees to pay all costs, and expenses in connection with damage to the Project and any other work or property damaged as a result of defects in the Work of this Agreement.

Article 11. The Subcontractor will provide 100% performance and payment bonds, such bonds to be provided by a Corporate Surety qualified to do business in the State in which the Project is located and acceptable to the Contractor.

Article 12. This Agreement constitutes the entire Agreement between the parties hereto and supercedes any prior proposals, quotations, understandings, representations or agreements, oral or written. This Agreement may not be changed in any way except as herein provided and no term or provision of this Agreement may be waived by any employee of the Contractor except in writing signed by its duly authorized officer.

Article 13. The Work under this Agreement may at any time be terminated in whole or in part by the Contractor. The termination shall be effective upon delivery to the Subcontractor of a Notice of Termination specifying the date upon which the termination becomes effective and the extent to which the Work is terminated. The Subcontractor shall not be entitled to anticipated overhead or profit on work unperformed or on equipment or materials unfurnished or any costs or damages due to a loss of business.

```
(If there are special conditions such as equal employment, other
federal acts, etc. involved for the project, they can be added
here as Article 14.)
```

CONTRACTOR: THE VOLPE CONSTRUCTION CO., INC.

By: _____

SUBCONTRACTOR: _____

By: _____

9/81 M

6

**AGREEMENT
BETWEEN OWNER
AND PROJECT ADMINISTRATOR**

(PROJECT ADMINISTRATOR AWARDS CONTRACTS)

(Cost Plus a Fee)

AGREEMENT

Made this   day of   in the year of Nineteen
Hundred and Ninety

BETWEEN

                              the Owner, and

                              the Project Administrator

For services in connection with the following described Project:

The Owner and the Project Administrator agree as set forth below:

1

ARTICLE 1

THE CONSTRUCTION TEAM AND EXTENT OF AGREEMENT

1.1   The Project Administrator accepts the relationship of trust and confidence established between the Project Administrator and the Owner by this Agreement.   The Project Administrator covenants with the Owner to furnish the Project Administrator's skill and judgment   in furthering the interests of the Owner.   The Project Administrator agrees to furnish construction administration and management services and to use the Project Administrator's best efforts to complete the Project in an expeditious and economical manner consistent with the interest of the Owner.   Nothing contained in this Agreement shall be deemed to require or authorize the Project Administrator to provide any service which could constitute the rendering of professional services. The reviews, recommendations, and advice to be furnished by the Project Administrator under this Agreement shall not be deemed to be warranties or guarantees or constitute performance of professional services.   The recommendation by the Project Administrator shall only be considered as a recommendation which is subject to the review and approval of the Owner.   The Project Administrator shall not be deemed to warrant the plans or design of the Architect/Engineer, or any other Consultants.

1.2   EXTENT OF AGREEMENT.   This Agreement represents the entire agreement between the Owner and the Project Administrator and supersede all prior negotiations, representations or agreements.   This Agreement may be amended only by written instrument signed by both the Owner and the Project Administrator.

1.3   DEFINED TERMS.   Terms used in this Agreement which are defined in Article 11 in the Contract Documents shall have the meanings designated in these Contract Documents.

ARTICLE 2

PROJECT ADMINISTRATOR'S SERVICES

2.1   PRECONSTRUCTION PHASE SERVICES.

.2.1.1   COMMENCEMENT OF PRECONSTRUCTION PHASE. The "Preconstruction Phase" will commence on or about _____.

2.1.2   The Project Administrator shall retain, with the Owner's approval, the Architect/Engineers for design and to prepare construction documents for the Work.   The Architect/Engineer's services, duties and responsibilities as described in the Agreement between the Project Administrator and the Architect/Engineer shall not be modified without written notification to the Owner.

2

2.1.3   The Project Administrator shall cause the Architect/Engineer and Consultants to carry such Professional Liability Insurance as the Owner and Project Administrator agree to adequately protect the Architect/Engineer against claims which may arise out of its professional liability.   Any losses in excess of the agreed amount of professional liability insurance will be paid by the Owner.   Such insurance policy shall be available for the inspection of the Project Administrator, who shall receive at least 30 days' notice prior to cancellation.   To the fullest extent permitted by law, the Architect/Engineer shall indemnify and hold harmless the Owner, Project Administrator and its Contractors, agents and employees from and against any and all loss, expense or damage (including, but not limited to, attorney's fees) arising out of the professional liability of the Architect/Engineer, the Architect/Engineer's consultants, and the agents and employees of any of them.

2.1.4   The Project Administrator, in its agreement with the Architect/ Engineer, shall include a provision requiring the Architect/Engineer to agree to consolidation of arbitration with the Owner and Project Administrator.

2.1.5   CONSULTATION DURING PROJECT DEVELOPMENT.   The Project Administrator shall schedule and attend regular meetings with the Architect/- Engineer during the development of conceptual and preliminary design to advise on site use and improvements, selection of materials, building systems and equipment.   The Project Administrator shall provide recommendations on construction feasibility, availability of materials and labor, time requirements for installation and construction, and factors related to cost including costs of alternative designs or materials and possible economies without assuming any of the Architects/Engineers or Consultants responsibility for design.

2.1.6   SCHEDULING.   The Project Administrator shall develop a Preliminary Time Schedule for Owner's approval that coordinates and integrates the Architect/Engineer's design efforts with construction schedules.

2.1.7   COST ESTIMATE.   As soon as the major requirements have been identified for any phase of the Work the Architect/Engineer or Contractors involved will prepare Costs Estimates for the construction of the Work based on such requirements. When the Schematic Design Phase is completed a Cost Estimate based on a quantity takeoff will be prepared by the Architect/Engineer and or Contractors involved.   This Cost Estimate will be updated as the plans and specifications are further developed.   All Cost Estimates will be submitted to the Project Administrator for his review.   The Project Administrator shall continue to review the Cost Estimates as the development of the Drawings and Specifications proceed.   The Project Administrator shall advise the Owner if it appears that the Budget will not be met and make recommendations for corrective action.   It is expressly understood that the Cost Estimate may be based upon incomplete design documents, is solely for the purpose of aiding in feasibility decisions by the Owner, and is not to be interpreted in any way as a guarantee of costs.

2.1.8   OWNER'S CONSULTANTS.   The Project Administrator shall, with the Owner's approval, select and retain professional services of a registered land surveyor, testing laboratories and special consultants, and if required, coordinate these services without, in either event, assuming any responsibility or liability for these consultants.

3

2.1.9   REVIEW OF CONSTRUCTION CONTRACT DOCUMENTS. The Project Administrator shall carefully review the Construction Contract Documents and shall at once report to the Owner any error, inconsistency, or omission the Project Administrator may reasonably discover.   The Project Administrator shall not be liable to the Owner or the Architect/Engineer for any damage resulting from any such errors, inconsistencies, or omissions.

2.1.9.1 Unless otherwise provided in the Construction Contract Documents, it is not the responsibility of the Project Administrator to make certain that the Construction Contract Documents are in accordance with applicable laws, statutes, building codes, and regulations.   If the Project Administrator observes that any of the Construction Contract Documents are at variance therewith in any respect, the Project Administrator shall promptly notify the Architect/-Engineer in writing, and the Architect/Engineer shall be responsible for making all changes necessary to correct such variance.

2.1.9.2 The Project Administrator shall make recommendations to the Architect/Engineer regarding the division of work in the Construction Draw-ings and Specifications to facilitate the bidding and awarding of   Construction Contracts, and. if necessary, allowing for phased construction, taking into consideration such factors as time of performance, availability of labor, overlapping trade jurisdictions, and provisions for temporary facilities.

2.1.9.3 The Project Administrator shall review the Construction Drawings and Specifications with the Architect/Engineer in an effort to eliminate areas of conflict and overlapping in the work to be performed by the various Construction Contractors.

2.1.10 PREQUALIFICATION.   The Project Administrator shall prepare prequalification criteria for bidders.

2.1.11 PRECONSTRUCTION PROCUREMENT. The Project Administrator shall recommend for purchase and expedite the procurement of long-lead items to ensure their delivery by the required dates.

2.1.12 INSURANCE.   The Project Administrator will purchase and maintain property insurance for the Project as provided in Article 10 herein and which costs are reimbursable under Article 7.   This insurance will be for limits with deductibles approved by the Owner.   All losses not covered by insurance will be a cost of the Work.

2.2   CONSTRUCTION PHASE SERVICES.

2.2.1   COMMENCEMENT OF CONSTRUCTION PHASE. The "Construction Phase" is estimated to start on _____.   The actual Construction Phase start will be the date established in the Notice to Proceed.

2.2.2   PROJECT TIME SCHEDULE.   The Project Administrator shall update the Time Schedule incorporating a detailed schedule for the construction operations, including realistic activity sequences and durations, allocation of labor and materials, processing of shop drawings and samples, and delivery of products requiring long lead-time procurement.   The Time Schedule shall include the Owner's occupancy requirements showing portions of the Work having occupancy priority.

2.2.3   PERMITS AND FEES.   The Project Administrator shall assist the Owner in obtaining all Building Permits and Special Permits for permanent improvements, excluding permits for inspection or temporary facilities required to be obtained directly by the various Contractors.   The Project Administrator shall assist in obtaining approvals from all the authorities having jurisdiction.

2.2.4   AFFIRMATIVE ACTION.   The Project Administrator shall determine applicable requirements for Affirmative Action Programs for inclusion in the bidding documents.

2.2.5   CONTRACTS.   The Project Administrator shall develop Contractor interest in the Work and as the Drawings and Specifications are completed, take competitive bids on the Work from prospective Construction or Turnkey Contractors.   After analyzing the bids and receipt of approval from the Owner, the Project Administrator shall award the Construction or Turnkey Contracts.

2.2.6   SITE ORGANIZATION.   The Project Administrator shall maintain a competent staff at the Project site to coordinate and oversee the   general direction of the Project and progress of the Construction or Turnkey Contractors.

.1   The Project Administrator shall establish on-site organization and lines of authority in order to carry out the overall plans of the Project.   The Project Administrator shall employ competent a Project Manager and necessary assistants who shall coordinate and provide general direction of the Work.   The Project Manager shall be satisfactory to the Owner, and shall not be changed except with the consent of the Owner, unless the Project Manager proves to be unsatisfactory to the Project Administrator or ceases to be in the Project Administrator's employ.   The Project Manager shall represent the Project Administrator and all communications given to the Project Manager shall be as binding as if given to the Project Administrator.   All communications shall be confirmed in writing.

.2   The Project Administrator shall establish procedures for coordination among the Owner, Architect/Engineer, and Contractors with respect to all aspects of the Work and implement such procedures.

.3   The Project Administrator shall schedule and conduct Progress Meetings at which   Contractors, Owner, Architect/Engineer and Consultants, if needed, can discuss jointly such matters as procedures, progress, problems and scheduling.

5

2.2.7 HAZARDOUS MATERIALS

2.2.7.1   The Project Administrator, under the direction of the Owner, shall have such tests done as are necessary to determine the true nature of such suspected material.   If such material is determined to be asbestos, polychlorinated biphenol (PCB) or other hazardous material, the Project Administrator under the direction of the Owner, shall have the responsibility for taking such action as is necessary to remove the hazardous material or to otherwise render it harmless consistent with statutes and/or regulations applicable to such material in the location of the Project.

2.2.7.2   If in fact the material is asbestos, PCB or other hazardous material the work shall stop and until it has been rendered harmless, the Work in the affected area shall not thereafter be resumed except by written agreement of the Owner and Project Administrator. The Work in the affected area shall be resumed in the absence of asbestos, PCB or other hazardous material, or when it has been rendered harmless, by written agreement of the Owner and Project Administrator.

2.2.7.3   The Project Administrator shall not be required, pursuant to Article 9 or such other provision of the Contract Documents, to perform without consenting any work relating to asbestos, PCB or other hazardous material.

2.2.7.4   To the fullest extent permitted by law, the Owner shall indemnify and hold harmless the Project Administrator, Architect/Engineer, Architect/-Engineer's consultants, Contractors and agents and employees of any of them from and against claims, damages, losses and expenses, including but not limited to attorneys' fees, arising out of or resulting from performance of the Work in the affected area if, in fact, the material is asbestos or PCB or other hazardous material and has not been rendered harmless, provided that such claim, damage, loss or expense is attributable to bodily injury, sickness, disease or death, or to injury to or destruction of tangible property (other than the Work itself) including loss of use resulting therefrom.   Such obligation shall not be construed to negate, abridge, or reduce other rights or obligations of indemnity which would otherwise exist as to a party or person described in this Subparagraph 2.2.7.4.

2.2.8   DOCUMENT INTERPRETATION. The Project Administrator shall refer all questions for interpretation of the Construction Documents prepared by the Architect/Engineer to the Architect/Engineer who shall provide such interpretation.   His decision as regards to esthetics shall be final.

2.2.9   SHOP DRAWINGS, PRODUCT DATA AND SAMPLES. The Project Administrator shall, in collaboration with the Architect/Engineer, establish and implement procedures for expediting the processing and approval of Shop Drawings, Product Data and Samples.

2.2.10 PHYSICAL CONSTRUCTION. The Project Administrator shall provide all supervision, labor, materials, construction equipment, tools and subcontract items which are necessary for the completion of the Work which is not provided by either the Contractors or the Owner.   To the extent that such Project Administrator performs any of the Work with his own forces, the

Project Administrator shall, with respect to such work, perform the work in accordance with the Construction or Turnkey Contract Documents.

2.2.11 DESIGNATED REPRESENTATIVE. The Project Administrator shall designate a representative who shall be fully acquainted with the Project and have authority to approve Change Orders, render decisions promptly, and furnish information expeditiously on behalf of the Project Administrator.

2.2.12 REVIEW OF WORK. The Project Administrator shall review the work of Construction or Turnkey Contractors for defects and deficiencies in the Work without assuming any of the Architect/Engineer's responsibilities for inspection.

2.2.13 PROGRESS REPORTS. The Project Administrator shall record the progress of the Work. The Project Administrator shall submit written progress reports to the Owner including information on the Construction or Turnkey Contractors' work, and the percentage of completion. The Project Administrator shall keep a daily log available to the Owner.

2.2.14 COST CONTROL. The Project Administrator shall develop and monitor an effective system of cost control for the Project. The Project Administrator shall review and if necessary cause the estimate to be revised and refine the initially approved Cost Estimate, incorporate approved changes as they occur, and develop Cash Flow Reports and forecasts as needed. The Project Administrator shall identify variances between actual and budgeted or estimated costs and advise Owner whenever projected cost exceeds budgets or estimates.

.1   The Project Administrator shall maintain cost accounting records on authorized Work performed under unit costs, actual costs for labor and material, or other basis requiring accounting records.

2.2.14 CHANGE ORDERS. The Project Administrator shall develop and implement a system for the preparation, review and processing of Change Orders. The Project Administrator shall recommend necessary or desirable changes to the Owner, review requests for changes, submit recommendations to the Owner, and assist in negotiating Change Orders.

2.2.15 PAYMENTS TO CONSTRUCTION OR TURNKEY CONTRACTORS. The Project Administrator shall develop and implement a procedure for the review, processing and payment of applications by Construction or Turnkey Contractors for progress and final payments.

2.2.16 SUBSTANTIAL COMPLETION. The Project Administrator shall determine Substantial Completion of the Work or designated portions thereof and the Architect/Engineer shall prepare a list of incomplete or unsatisfactory items and schedule for their completion.

2.2.17 FINAL COMPLETION. The Project Administrator shall determine final completion of the various phases of the Work and provide written notice to the Architect/Engineer that the Work is ready for final inspection. The Architect/Engineer shall secure and transmit to the Project Administrator all required guarantees, affidavits, releases, bonds and waivers. The Project

7

Administrator shall turn over all keys, manuals, record drawings and maintenance stocks to the Owner who shall acknowledge receipt of same.

2.3   ADDITIONAL SERVICES.

2.3.1   At the request of, and as a service to the Owner, the Project Administrator will provide the following, or other, additional services upon written agreement between the Owner and Project Administrator defining the extent of such additional services and the amount and manner in which the Project Administrator will be compensated for such additional services. Services provided under this paragraph 2.3 are not guaranteed or warranteed by the Project Administrator.

2.3.1.1 Services related to investigation, appraisals or valuations of existing conditions, facilities or equipment, or verifying the accuracy of existing drawings or other Owner-furnished information.

2.3.1.2 Services related to Owner-furnished equipment, furniture and furnishings which are not a part of this Agreement.

2.3.1.3 Obtaining or training maintenance personnel or negotiating maintenance service contracts.

ARTICLE 3

OWNER'S RESPONSIBILITIES

3.1   The Owner shall furnish for the site of the Project all necessary surveys describing the physical characteristics, soil reports and subsurface investigations, legal limitations and utility locations, and a legal description of the site. It is further understood that the Project Administrator can rely on the accuracy of these surveys and reports.

3.2   The Owner will secure and pay for necessary approvals, easements, assessments and charges required for the construction, use, or occupancy of permanent structures or for permanent changes in existing facilities, required by governing authorities for the construction of the Project.

3.3   The Owner shall furnish such legal services as may be necessary for providing the items set forth in Paragraph 3.2.

3.4   The Owner shall promptly notify and furnish information to the Project Administrator of any change in the Owner's ownership interest in the Project.

3.5   The Owner shall designate a representative who shall be fully acquainted with the Project and have authority to provide all decisions required of the Owner such as, but not limited to, approval of Cost Estimates, Requisitions, issuing Change Orders, render decisions promptly and furnish information expeditiously.

3.6   If the Owner becomes aware of any fault or defect in the Work or nonconformance with the Contract Documents, the Owner shall give prompt written notice thereof to the Project Administrator.

8

3.7   The Owner shall furnish, prior to commencing the Project and at such future times as may be requested by the Project Administrator, reasonable evidence satisfactory to the Project Administrator that sufficient funds are available and committed for the entire cost of the Project. Unless such reasonable evidence is furnished, the Project Administrator is not required to commence or continue any Work, or may, if such evidence is not presented within a reasonable time, stop the Work upon 15 days' notice to the Owner. The failure of the Project Administrator to insist upon the Owner providing this evidence at any one time shall not be a waiver of the Owner's obligation to make payments pursuant to this Agreement,, nor shall it be a waiver of the Project Administrator's right to request or insist that such evidence be provided at a later date.

3.8   The Owner shall communicate with the Construction or Turnkey Contractors and the Architect/Engineer only through the Project Administrator.

3.9   SEPARATE CONTRACTORS.

3.9.1 If the Owner performs work related to the Project with his own forces or awards separate contracts in connection with other portions of the Project or other work on the site, the Owner will provide for the coordination of the work of his own forces and of each separate contractor with the Work of the Project Administrator.

3.9.2 If the Project Administrator claims that delay or additional cost is involved because of the actions or conduct of the Owner's own forces or by his separate contractors performing work on the Project the Project Administrator shall make such claim as provided elsewhere in the Contract Documents.

ARTICLE 4

CONSTRUCTION CONTRACTS

4.1   All portions of the Work shall be performed under Construction or Turnkey Contracts approved by the Owner. The Project Administrator shall request and receive proposals from Construction or Turnkey Contractors. Based on the proposals received, the Project Administrator will make a recommendation for the award of the Construction or Turnkey Contracts. The Project Administrator's recommendation will be reviewed with the Owner. The Construction or Turnkey Contract will be awarded after approval by the Owner.

4.2   If the Owner refuses to accept a Construction or Turnkey Contractor recommended by the Project Administrator, the Project Administrator shall recommend an acceptable substitute. The Cost Estimate and the Time Schedule, and the Date of Substantial Completion, if affected, shall be adjusted by the difference in cost and construction time resulting from such substitution.

4.3   Construction or Turnkey Contracts will be between the Project Administrator and the Construction or Turnkey Contractors.

4.4   The Construction or Turnkey Contractors will be required to provide Payment and Performance Bonds if deemed appropriate by the Project Administrator or required by the Owner.

ARTICLE 5

SCHEDULE

5.2   If the progress of the Project is delayed at any time by any act or neglect of the Owner or by any of his employees, or by any separate contractor employed by the Owner, or by Changes in the Work, or by labor disputes, fire, unusual delay in transportation, adverse weather conditions not reasonably anticipated, unavoidable casualties or any causes beyond the Project Administrator's control, or by delay authorized by the Owner pending arbitration, the Date for Substantial Completion of the Work shall be extended by Change Order for a reasonable length of time including any additional costs incurred by the Project Administrator.

ARTICLE 6

PROJECT ADMINISTRATOR'S FEE

6.1   In consideration of the performance of this Agreement, the Owner agrees to pay the Project Administrator in current funds as compensation for the Project Administrator's services a Fee as set forth in Subparagraphs 6.1.1.

6.1.1   For the services performed during the Preconstruction and Construction Phases of the project, the Project Administrator shall be paid a Fee of _____ _____(_____%) of the total Project cost.   The Fee shall be paid monthly based on the value of the Project then completed.   (The reimbursable costs incurred by the Project Manager shall also be   paid monthly as per his requisition.)

6.2   Adjustments in the Project Administrator's Fee shall be made as follows:

6.2.1   For Changes in the Work as provided in Article 8, the Project Administrator's Fee shall be adjusted as follows:

The same fees and conditions as set forth in paragraph 6.1.1

6.2.2   The Project Administrator shall be paid an additional Project Administrator's Fee in the same portion as was set forth in Subparagraph 6.2.1 if the Project Administrator is placed in charge of the reconstruction of any insured or uninsured loss.

6.3   Included in the Project Administrator's Fee are the following:

6.3.1   Salaries or other compensation of the Project Administrator's employees at its principal office and branch offices, except employees listed in Subparagraph   7.2.2.

6.3.2   General operating expenses of the Project Administrator's principal and branch offices, other than the field office, except as included in Article 7.

1 0

6.3.3 Any part of the Project Administrator's capital expenses, including interest on the Project Administrator's capital employed for the Work.

ARTICLE 7

COST OF THE WORK

7.1 The term Cost Of The Work shall mean all costs necessarily incurred by the Project Administrator during either the Preconstruction or Construction Phase in the performance of this Agreement. Such costs shall include the items set forth below in this Article and are reimbursable to the Project Administrator.

7.1.1 The Owner agrees to pay the Project Administrator for the Cost Of The Work as defined in this Article 7. Such payment shall be in addition to the Project Administrator's Fee stipulated in Article 6.

7.2 COST ITEMS.

7.2.1 Wages paid for construction workers in the direct employ of the Project Administrator in the performance of the Work under applicable collective bargaining agreements, or under a salary or wage schedule agreed upon by the Owner and Project Administrator, and including employee benefits, if any, as may be payable with respect thereto.

7.2.2 Wages or salaries of the Project Administrator's supervisory and administrative employees when stationed at the field office, in whatever capacity employed, employees engaged in expediting the production or transportation of materials and equipment, and such employees in the main or branch office performing the functions listed below:

> Project Managing
> Estimating
> Purchasing

and employees performing the Preconstruction Phase Services wherever located, as listed below:
> Project Managers
> Estimators

7.2.3 Costs paid or incurred by the Project Administrator for taxes, insurance, contributions, assessments, and benefits required by law or collective bargaining agreements and, for personnel not covered by such agreements, customary benefits such as sick leave, medical and health benefits, holidays, vacation, pensions and or profit sharing, provided such costs are based on wages and salaries included in the Cost Of The Work under Subparagraphs 7.2.1 and 7.2.2.

7.2.4 Cost of transportation, traveling, moving and hotel expenses of the Project Administrator or of the Project Administrator's officers or employees incurred in discharge of duties connected with the Work.

1 1

7.2.5   Cost of all materials, supplies and equipment incorporated in the Project, including costs of transportation and storage thereof.

7.2.6   Payments made by the Project Administrator to Construction or Turnkey Contractors for their work and costs of the Architect/Engineer's and Consultant's Contracts.

7.2.7   Cost, including transportation and maintenance of all materials, supplies, equipment, hand tools not owned by the workers which are employed or consumed in the performance of the Project, and temporary facilities (including trailers, office supplies, computers, surveying equipment and other related items necessary for overseeing the Project).

7.2.8   Rental charges of all necessary machinery and equipment, exclusive of hand tools, used at the site, whether rented from the Project Administrator or others, including insurance, installation, repairs and replacements, dismantling, removal, lubrication, fuel, transportation and delivery costs thereof, at rental charges consistent with those prevailing in the area.

7.2.9   Cost of the premiums attributable to this Agreement for all insurance and bonds which the Project Administrator is required to maintain pursuant to the Contract Documents or is deemed necessary by the Project Administrator including Architect/Engineers and or Consultants errors and omissions insurance.

7.2.10   Sales, use, gross receipts or similar taxes related to the Work imposed by any governmental authority, and for which the Project Administrator or the Project is liable.

7.2.11   Permit fees, licenses, tests, royalties, damages for infringement of patents and costs of defending suits therefor, and deposits lost for causes other than the Project Administrator's negligence. If royalties or losses and damages, including attorneys' fees and cost of defense are incurred, which arise from a particular design, process, or the product of particular manufacturer or manufacturers specified by the Owner or Architect/-Engineer, and the Project Administrator has no reason to believe there will be infringement of patent rights, such royalties, losses and damages including attorneys' fees and costs of defense shall be deemed a cost of the Project.

7.2.12   Losses, expenses or damages to the extent not compensated by insurance or otherwise (including settlement made with the written approval of the Owner). The amount of the deductible limits of insurance covering such loss shall be a Cost Of The Work.

7.2.13   The cost of correcting defective or rejected work and the cost of warranty work as to any portion of the Work performed by Project Administrator's own forces, and as to a Construction or Turnkey Contractor's Work if the Contractor is bankrupt, unable to perform, or no longer in business. If the Construction or Turnkey Contractor's Work is covered by a Performance and Payment Bond the work done shall be paid as a cost of the work and the recovery of funds shall be paid to the Owner when recovered.

1 2

7.2.14   Minor expenses including but not limited to telegrams, long-distance telephone calls, telephone service , expressage, progress photographs, printing, reproduction, and similar petty cash items in connection with the Work.

7.2.15   Cost incurred due to an emergency affecting the safety of persons or property.

7.2.16   Cost of data processing services required in the performance of the Work including C P M consultants.

7.2.17   Legal costs reasonably and properly resulting from prosecution of the Project for the Owner or the enforcement of Construction or Turnkey Contracts.

7.2.18   All costs directly incurred in the performance of the Work and not included in the Project Administrator's Fee as set forth in Paragraph 6.3.

7.3   COSTS ASSOCIATED WITH CHANGES.

7.3.1   The costs associated with Changes include all of the items set forth in this Article 7.

ARTICLE 8

CHANGES IN THE WORK

8.1   The Owner, without invalidating this Agreement, may order Changes in the Work within the general scope of this Agreement consisting of additions, deletions or other revisions, with the Project Administrator's Fee, supervision, expenses related to time and additional work and the Date of Substantial Completion being adjusted in accordance with the other provisions of this Agreement.   All such Changes in the Work shall be authorized by Change Order and shall be performed under the applicable conditions of the Contract Documents.

8.2   EMERGENCIES.

8.2.1   In any emergency affecting the safety of persons or property, the Project Administrator shall act, at his discretion, to prevent damage, injury or loss from hurricanes, floods, storms, high winds, etc..   Any increase in the Project Administrator's Fee, costs of the work or extension of time claimed by the Project Administrator on account of emergency work shall be determined as provided in this Article 8.

8.3   CLAIMS FOR ADJUSTMENT IN FEE OR TIME.

8.3.1   If the Project Administrator wishes to make a claim for an adjustment in the Project Administrator's Fee, or a revision of the Time Schedule, the Project Administrator shall give the Owner written notice thereof within a reasonable time after the observance of the event giving rise to such claim. This notice shall be given by the Project Administrator before proceeding to execute the portion of the Work, if any, which is the subject matter of the claim, except in an emergency endangering life or property in which case the Project Administrator shall act, at the Project Administrator's discretion, to prevent threatened damage, injury or loss.   Claims arising from delay shall be made within a reasonable time after the delay.   No such claim shall be valid unless so made.   If the Owner and the Project Administrator cannot agree on the amount of the adjustment in the Project Administrator's Fee or the Date of Substantial Completion, it shall be determined pursuant to the provisions of Article 14.   Any change in the Project Administrator's Fee or the Time Schedule resulting from such claim shall be authorized by Change Order.

ARTICLE 9

PAYMENTS TO THE PROJECT ADMINISTRATOR

9.1   The Project Administrator shall submit monthly to the Owner a statement, sworn to if required, showing in detail all moneys paid out, costs accumulated or costs incurred on account of the Cost Of The Work during the previous month less retainage set forth in any Construction or Turnkey Contract and the amount of the Project Administrator's Fee due as provided in Article 6 and cost of the Project Administrator's as provided for in Article 7. Payment by the Owner to the Project Administrator of the statement amount shall be made within ten (10) days after it is submitted.

9.2   Final payment constituting the unpaid balance of the Cost of the Work and the Project Administrator's Fee shall be due and payable for each phase of the Project when it is delivered to the Owner, ready for beneficial occupancy, or when the Owner occupies the Project, whichever event first occurs, provided that the Work be then substantially completed and this Agreement substantially performed.   If there should remain minor items to be completed, the Architect/Engineer shall deliver, in writing, the Construction or Turnkey Contractors unconditional promise to complete said items within a reasonable time there-after to the Project Administrator.   The Project Administrator may retain a sum equal to 150% of the estimated cost of completing any unfinished items like-wise listed separately.   Thereafter, the Owner shall pay to the Construction or Turnkey Contractors through the Project Administrator, monthly, the amount retained for incomplete items as each of said items is completed.

9.3   The Project Administrator shall promptly pay all the amounts due the Construction or Turnkey Contractors or other persons with whom the Project Adminis-trator has a contract upon receipt of any payment from the Owner, the application for which includes amounts due such   Contractors or other persons.   Before issuance of final payment, the Project Administrator shall submit satisfactory evidence that all payrolls, material bills and other indebtedness connected with the Project have been paid or otherwise satisfied.

9.4   If the Owner should fail to pay the Project Administrator within seven (7) days after the time the payment of any amount becomes due, then the Project Administrator may, upon seven (7) days written notice to the Owner, stop the Work until payment of the amount owing has been received.

9.5   Payments due and unpaid shall bear interest from the date payment is due at two percentage points in excess of the Prime   Rate prevailing in the area .

9.6   If, subsequent to final payment, the Project Administrator incurs costs described in Article 7 to correct defective or nonconforming work or to perform warranty work, the Owner shall reimburse the Project Administrator for such costs and the Project Administrator's Fee applicable thereto on the same basis as if such costs had been incurred prior to final payment.

ARTICLE 10

INSURANCE AND WAIVER OF SUBROGATION

10.1   INSURANCE.

10.1.1 The Project Administrator shall purchase and maintain such insurance coverages as set forth in this Article 8 and such other insurance as he may deem necessary.   The Comprehensive General Liability Insurance and Automobile Liability Insurance, shall be written for not less than the limits of liability as listed in Schedule "A" attached hereto

10.2   OWNER'S LIABILITY INSURANCE.

10.2.1 The Owner shall purchase and maintain the Owner's own liability insurance, including insurance for damage to adjacent property, and, at the Owner's option, may purchase and maintain such insurance as will protect the Owner against claims which may arise from operations under this Agreement.

10.3   PROPERTY INSURANCE.

10.3.1 The Project Administrator will purchase, as a reimbursable item, and maintain property insurance upon the entire Project.   This policy and its deductions will be approved by the Owner.

10.4   WAIVER OF SUBROGATION.

10.4.1   The Owner and Project Administrator waive all rights against each other, the Architect/Engineer, Construction or Turnkey Contractors and their Trade Contractors.

15

ARTICLE 11

DEFINITIONS

11.1   THE PROJECT. The "Project" is the total design and construction of which the Work performed under other contracts may be the whole or a part.

11.2   THE WORK. The term "Work" means the Construction and Design Phase Services to be performed under the Project Management Agreement. The Work may constitute the whole or part of the Project.

11.3   DAY. The term "Day" shall mean calendar day unless otherwise specifically designated.

11.4   OWNER. The "Owner" is the person or entity identified as such in the Agreement between the Owner and Project Administrator. The term "Owner" means the Owner or the Owners authorized representative.

11.5   PROJECT ADMINISTRATOR. The Project Administrator is the person or entity who has entered into the Agreement with the Owner to serve as Project Administrator.   The term "Project Administrator" means the Project Administrator or the Project Administrators authorized representative.

11.6   CONSTRUCTION OR TURNKEY CONTRACTORS. A Construction or Turnkey Contractor is a person or entity who performs a portion of the Work pursuant to an agreement with the Project Administrator and who is identified as a Construction or Turnkey Contractor in its agreement with the Project Administrator.   The term "Construction or Turnkey Contractor" means the Construction or Turnkey Contractor or an authorized representative.

ARTICLE 12

TERMINATION OF THE AGREEMENT AND OWNER'S
RIGHT TO PERFORM PROJECT ADMINISTRATOR'S OBLIGATIONS

12.1   If the Work, in whole or substantial part, is stopped for a period of thirty days under an order of any court or other public authority having jurisdiction, or as a result of an act of government, such as a declaration of a national emergency making materials unavailable, through no act or fault of the Project Administrator, or if the Work should be stopped for a period of thirty days by the Project Administrator for the Owner's failure to make payment thereon, then the Project Administrator may, upon seven days' written notice to the Owner, terminate this Agreement and recover from the Owner payment for all Work executed, the Project Administrator's Fee earned to date, and for any proven loss sustained upon any materials, equipment, tools, construction equipment and machinery, cancellation charges on existing obligations of the Project Administrator, damages, and a reasonable profit.

16

12.2   OWNER'S RIGHT TO PERFORM PROJECT ADMINISTRATOR'S OBLIGATIONS AND TERMINATION BY THE OWNER FOR CAUSE.

12.2.1 If the Project Administrator persistently or repeatedly refuses or fails, except in cases for which extension of time is provided, to supply proper supervision of the project or if the Project Administrator fails to make proper payment to Contractors or persistently disregards laws, ordinances, rules, regulations or orders of any public authority having jurisdiction or otherwise is guilty of a substantial violation of a provision of the Agreement, the Owner may, without prejudice to any right or remedy and after giving the Project Administrator seven days' written notice, during which period the Project Administrator fails to commence to cure the violation, terminate the employment of the Project Administrator and take possession of the site and may finish the Work by whatever reasonable method the Owner may deem expedient.   In such case, the Project Administrator shall not be entitled to receive any further payment, due the Project Administrator until the Project is finished.

ARTICLE 13

ASSIGNMENT

13.1   Neither the Owner nor the Project Administrator shall assign their interest in this Agreement without the written consent of the other except as to the assignment of proceeds or a collateral assignment by the Owner to the construction lender providing funds for the Project.

ARTICLE 14

ARBITRATION

14.1   All claims, disputes and other matters in question arising out of, or relating to, this Agreement or the breach thereof, except with respect to the Architect/Engineer's decision on matters relating to artistic effect, and except for claims which have been waived by the making or acceptance of final payment, shall be decided by arbitration in accordance with the Construction Industry Arbitration Rules of the American Arbitration Association Industry Arbitration Rules of the American Arbitration Association then obtaining unless the parties mutually agree otherwise.   This Agreement to arbitrate shall be specifically enforceable under the Federal Arbitration Act.

14.2   If a dispute arises out of or relates to this Agreement, or the breach thereof, the parties shall endeavor to settle the dispute first through direct discussions.   If the dispute cannot be settled through direct discussions, the parties shall endeavor to settle the dispute by mediation under the Construction Industry Mediation Rules of the American Arbitration Association.   Mediation shall commence within the time limits for arbitration stipulated in the Contract Documents.   The time limits for any subsequent arbitration shall be extended for the duration of the mediation process plus ten (10) days or as otherwise provided in the Contract Documents.   Issues to be mediated are subject to the exceptions in Paragraph 14.1 for arbitration.   The location of the mediation shall be the same as the location for arbitration identified in Paragraph 14.3.

14.3   Notice of the demand for arbitration shall be filed in writing with the other party to this Agreement and with the American Arbitration Association. The demand for arbitration shall be made within a reasonable time after the claim, dispute or other matter in question has arisen, and in no event shall it be made after the date when institution of legal or equitable proceedings based on such claim, dispute or other matters in question would be barred by the applicable statute of limitations.   The location of the arbitration proceeding shall be in the state in which the project is located or as otherwise mutually agreed to by the parties.

14.4   The award rendered by the arbitrators shall be final and judgment may be entered upon it in accordance with applicable law in any court having jurisdiction   thereof.

14.5   Unless otherwise agreed in writing, the Project Administrator shall carry on the Work and maintain the Date of Substantial Completion during any arbitration proceedings, and the Owner shall continue to make payments in accordance with this Agreement.

14.6   All claims which are related to or dependent upon each other shall be heard by the same arbitrator or arbitrators even though the parties are not the same, unless a specific contract provision prohibits such consolidation.

This Agreement executed the day and year first written above.

ATTEST:_____   OWNER:

                                         _____

ATTEST:_____   PROJECT ADMINISTRATOR:

                                         _____

SCHEDULE A

The Comprehensive General Liability Insurance and Automobile Liability Insurance, as required by Subparagraph 11.1.1 shall be written for not less than the limits or liability as follow:

a. Comprehensive General Liability

    1. Bodily Injury $_____ Each Occurrence

                                      $_____ Aggregate (Completed Operations)

    2. Property Damage $_____ Each Occurrence

                                        $_____ Aggregate

b. Comprehensive Automobile Liability

    1. Bodily Injury $_____ Each Person

                                        $_____ Each Occurrence

    2. Property Damage $_____ Each Occurrence

c. Umbrella Coverage $_____ Total Amount

2/12/90

19

ABOUT THE AUTHORS

S. Peter Volpe is President of the Volpe Construction Company of Malden, Massachusetts, and the author of CONSTRUCTION MANAGEMENT PRACTICE, published by John Wiley & Sons, Inc. in 1972. He is Past President and Life Director of the Associated General Contractors of America, and Past President of the Associated General Contractors of Massachusetts. He has also served as Chairman of the Building Division of the AGC of America. He is a Fellow of the Society of American Military Engineers, a Panelist with the American Arbitration Association, and a member of the Consulting Constructors' Council of America and The Moles. He also served in the Navy Seebees in both the Atlantic and Pacific theaters in World War II.

Peter J. Volpe is a Senior Staff Member of the Volpe Construction Company. He has been affiliated with the firm since his graduation from Merrimack College in 1971. He is a Past President and Director of the Society of American Military Engineers, Boston, and a past Chairman of the Crime Prevention Committee of the AGC of Massachusetts.

INDEX